Rüdiger Mach
Peter Petschek

# Visualization of Digital Terrain and Landscape Data

Rüdiger Mach
Peter Petschek

# Visualization of Digital Terrain and Landscape Data

## A Manual

With 264 Figures

Autodesk®

Springer

DIPL.-ING. RÜDIGER MACH

Lerchenberg 3
8046 Zurich
Switzerland
*E-mail: 3d_info@terrainviz.com*

PROF. DIPL.-ING. PETER PETSCHEK

HSR Rapperswil
Oberseestr. 10
8640 Rapperswil
Switzerland
*E-mail: peter.petschek@hsr.ch*

**Foreword by:**

DR. STEPHEN M. ERVIN

Harvard University
Graduate School of Design
48 Quincy St.
Cambridge MA 02138
USA

ISBN 978-3-642-06783-9 e-ISBN 978-3-540-30491-3

Original title: "Visualisierung digitaler Gelände- und Landschaftsdaten"



Springer is a part of Springer Science+Business Media
springer.com


Cover design: deblik, Berlin

# Dedication and Acknowledgement

For Christine, Wolf and Ruven (thank you all for just being there)
*Rüdiger Mach*

For Luciana and my parents
*Peter Petschek*

The list of the Kind, the Patient and the Helpful who have contributed to this book now includes the following very special persons: M. von Gadow (It is only thanks to Marga, that the English version came into existence), Prof. K. von Gadow, Dr. S. Ervin and Y. Maurer.

We are especially grateful for the support and encouragement of Christine (for correcting the horrible mistakes we made with the english language) and of course of the software producers, which where mentioned in this book:
P. Rummel (Autodesk), Dr. M. Beck (ViewTec Ltd.), F. Staudacher (Leica-Geosystems), D. McClure und J. Hervin (Digital Elements), C. Quintero (Itoo Software), E. May (Anark Corporation)

Special thanks to Dr. C. Witschel for his patience while waiting and waiting and …

# Foreword

This book reflects a profound change that has taken place in the practice of landscape architecture and planning in the past twenty years. Traditional modes of representation – pen, pencil, watercolor, marker, *et al* – have been supplanted by digital modeling and animation. This transformation is not just in the medium of representation, however; it is more than a substitution of one marking device for another, such as may have been the case in the past when, for example, mechanical pens with cartridges replaced pens with nibs that were filled by dipping. Even changes such as that had their impacts (as longer straighter lines, for example, or more precision in details became possible) on the interplay between designer, design medium, and designed artifact(s). The emergence of digital media as representational tools for designers has accompanied a transformation in the language of discourse in design and planning, in the very conception of the designed world we live in, and in the substance and role of the essential representations and abstractions used by planners and designers.

In the past, when 2D planar representations (drawings, usually on paper) served as the conventional means of communication for designers (both with themselves and with others), physical objects or arrangements in 3D were transformed into a series of lines in 2D (plans, sections, elevations, e.g.) by the draftsmen's and renderers' craft; those lines and images were then reconstructed by the viewer's "mind's eye", into a virtual construction, and evaluations and judgments were rendered, as to beauty, fit, function, etc. This stylized communication, still largely in place in many engineering, design, and construction contexts, depends upon a system of conventional signs and symbols, and in most cases, professional training, to achieve good communications (a high 'signal to noise ratio'). For many ordinary lay persons, the highly stylized 2D representations such as contour plans of terrain are virtually incomprehensible, and so the 'artists conception' , 'illustrative sketch', and 'perspective rendering' are widely used to convey basic 3D form, color, and texture so that the observer's mind's eye has less work to do. These 3D artists' renditions however, in spite of techniques that increasingly make them appear 'photorealistic', have a deserved reputation for bias and fictions, as the artists' subjectivity, agenda of persuasion or selling, and media-dependent distortions, often overwhelm the objective basis of the proposed plans.

With computer graphics and digital modeling techniques, the potential for distortion and variable interpretation is no less; but the technical details are much changed. The advent of affordable, complex, sophisticated software running on powerful desktop computers has brought the possibility of combining and exploring objectivity and subjectivity in new and exciting ways. With computer graphics visualizations, and techniques such as those detailed in this book, the hand of the artist (the designer, or renderer, or visualizer) is once removed from the artifact; and yet the final visualization is closer to an intermediate virtual representation. Whereas the watercolor renderer of old chose brush, paper, and color, and applied strokes and washes directly, in a deft moment that permitted little hesitation and with very little chance for re-doing or erasing, the modern digital artist has both more choices, more leisure, and a whole new universe of design possibilities, in the virtual world. These 3D visualizations have taken the place both of final artist's conceptions, and of designers' initial sketches and preliminary models. Their transforming role in both the 'praxis' and the 'poetics' of landscape architecture is undeniable.

Twenty-first century digital techniques allow for many subtle variations and transformations, for collage and assemblage, for un-doing, re-doing and for tracking multiple variations in the process of making virtual models and their visualizations. Modern digital artists 'sample' the real world in a variety of ways – much as modern musical artists sample music for re-mixing --, measuring, surveying, photographing, digitizing, generating, transforming, and combining data both from the objective world and from their designerly imaginations. The virtual models they create have a status all their own, independent either of the real-world conditions they may seek to illustrate, or the renderings and animations that will be produced from them. The images produced by software from these models can have an objective factual basis as accurate as their creator wishes; while at the same time, can contain fantastic and singular objects, phenomena, and effects in ways unimaginable just twenty years ago.

This book is a guide to this rich and evolving world of possibilities. The authors combine the speculative and adventurous curiosity of the academic and researcher with the rigorous and realistic experiences of the practitioner. Rüdiger Mach and Peter Petschek each bring their own particular viewpoint and experience, gained from both hypothetical exercises and reality-based work. Their facility with modern digital tools, both as science and as art, combined with their deep experience of built works and social processes, gives this book its unique flavor and value.

Although many of the basic techniques presented here have been known and used for some time, and some are now built-in to modern modeling and rendering software, their best uses and combinations are still being discovered. Putting them all together, providing a synthesis of both possibilities and opportunities with constraints and actual needs, is the great contribution of this volume. The discipline of landscape architecture provides a perfect testing ground for these approaches, covering as it does the broad range of human, technical and environmental concerns, from urban open space plans to residential gardens and regional forests and wilderness preserves. The essential subjects of landscape – terrain, vegetation, water, and the atmosphere that suffuses them – provide technical challenges to computer modeling, being as they are curved, irregular, fuzzy, fractal, fluid, ethereal, and dynamic. The technical approaches to modeling and visualizing these objects and phenomena are still emergent, but some techniques are well enough established that they can be and are effectively used today. The authors provide examples in the text of these techniques and projects across a range of scales and contexts, tied together with an orderly and structured approach which starts with systems of measurement and concludes with practical examples of workflow and organization of a real-world visualization project. Ethical concerns for creators of visualizations, as well as emergent techniques for collecting real time data and working with digital video are also covered.

This book is not so much a smorgasbord of digital techniques, as a well-conceived multi-course meal, with each part contributing to the overall effect.

Enjoy the feast!

Stephen M Ervin
Harvard Graduate School of Design
August 2005

# Contents

# About the Authors

**Rüdiger Mach** – Rüdiger Mach is a hydraulic engineer and has been working in the field of 2D and 3D CAD and computer graphics for almost two decades. He has specialized in the field of technical and scientific visualization and is writing articles and books about 3D visualization.
As a senior project manager of the ViewTec Ltd., he is responsible for interactive 3D visualization, scientific presentation of results and everything to do with the designing and communication of visual problem solving. He is a lecturer at the Hochschule für Technik und Wirtschaft in Karlsruhe, Germany and at the HSR Hochschule für Technik in Rapperswil, Switzerland.

**Peter Petschek** - Peter Petschek is a professor at the HSR Hochschule für Technik, Rapperswil, Switzerland, at the Department of Landscape Architecture. He lectures in the Bachelor and Master programs, and his courses are, among others, in CAD, DHM, and 3D computer visualization. Furthermore, he is head of the HSR post-graduate course in 3D computer visualization in planning and architecture. After his studies in landscape architecture at the Technical University of Berlin, Peter Petschek studied in the USA in the mid-80s (Master of Landscape Architecture). Here, he got to know CAD/GIS/ image editing. The implementation of information technology, especially in the field of digital elevation modeling in landscape architecture has been one of his main points of interest ever since.

# Introduction

*This chapter deals mainly with the basics. Questions to be answered are: what are digital terrain models, what is necessary in order to create accomplished landscape visualizations, who creates and who needs digital terrain models, and in which fields are digital terrain models used?*

**Fig. 1.** Isar Valley, near Bad Toelz, Germany

Terrain and landscape modeling is in itself a complicated theme. Many people want to be able to create geo-referenced digital terrain models, and of course, to be able to handle them interactively. This reminds us a little of the suddenly burgeoning enthusiasm with which GIS was celebrated at the beginning of the 90s. All of a sudden, everyone working in technical and scientific fields was talking about GIS, everyone wanted to use it and only few actually knew what GIS really is.

In the meantime, terrain models and the visualization of digital terrain data have become an important part of many plans, and plenty of them would surely have become much more expensive without visualization.

**hint Costs**

*By relaying designs in a well directed manner, visualization of geo data and laandscapes helps to avoid misunderstandings and thereby to reduce costs.*

Now, let us take a look at the exciting and extensive topic of terrain models and landscape data. At Wikipedia.org, "digital terrain model" is defined as follows:

*"A digital terrain model (abbrev. DTM) is a topographic model of the bare earth that can be manipulated by computer programs.*

*The data files contain the elevation data of the terrain in a digital format, which relates to a rectangular grid. Vegetation, buildings and other cultural features are removed digitally - leaving just the underlying terrain.*

*DTMs are used especially in civil engineering, geodesy & surveying, geophysics, geography and remote sensing. The main applications are:*

1. *Visualization of the terrain (Block diagrams etc.)*
2. *Calculation of project data or rock masses in civil engineering*
3. *Reduction (terrain correction) of gravity measurements (gravimetry, physical geodesy)*
4. *Terrain analyses in Cartography and Morphology*
5. *Rectification of airborne or satellite photos.*[1]

The same source defines the term "digital elevation model" as follows:

*"A digital elevation model (DEM) is a representation of the topography of the Earth or another surface in digital format, that is, by coordinates and numerical descriptions of altitude. DEMs are used often in geographic information systems. A DEM may or may not be accompanied by information about the ground cover. In contrast with topographical maps, the information is stored in a raster format. That is, the map will normally divide the area into rectangular pixels and store the elevation of each pixel."*

There are certainly many further descriptions and definitions, but let's leave it at that and get to the practical side of things.

[1] http://www.geoinformatik.uni-rostock.de

## Concepts

At the start of a visualization project, concepts are often used that – if they are not clearly defined – can lead to friction, misunderstandings and disappointment, or even disputes and arguments.

It's a good idea to attempt to reach a consensus about the language and the way in which certain concepts are used before starting to work on a project. The concepts, which most often cause difficulties due to differences in interpretation, are:

Photo realism
Realistic
Authenticity
Credibility

These concepts can be used in many different ways and their meaning depends on the context in which they are used.

In this way, a scene made in comic style can be credible and convincing, while at the same time not attempting to be realistic.

The successful realization of a scene for use in computer game demands for it to be credible and to leave the impression that it is authentic, but it does not have to be photo realistic.

However, if a scene in a movie is used as a matte painting, it must be photo realistic and not leave the shadow of a doubt about its authenticity.

Therefore, it's a good idea to define the requirements exactly with the client before starting to work on any project.

The term photo-realism is used too often and too quickly, although for many projects a realistic approach would suffice.

## Five Principles?

On looking at the long list of those who are seriously engaged in the topic of landscape visualization, Sheppard most certainly can be regarded as one of the pioneers. He lists five principles necessary for accomplished landscape visualization. These are:

**Representative character:** visualizations should represent typical or important aspects or conditions of the landscape.

**Exactness**: visualizations should simulate the factual or expected appearance of the landscape (at least for those factor, which will be evaluated)
**Optical Clarity:** details, components and content of the visualization should be clearly identifiable.
**Interest:** the visualization should arouse the interest of the viewer and hold their attention for as long as possible.
**Legitimacy:** the visualization should be justifiable and its level of exactness should be verifiable.

To be exact, one can add another important point to Sheppard's list, namely interactivity (e.g. mass-market web-based 3D visualizations such as Google Earth or TerrainView-Globe).

For the world of visualization, it is also of utmost importance that both the viewer and the visualizer are fascinated by visualizations. This does not make visualization great, but it greatly adds to the possibility that the visualization will be accepted by the viewer if the viewer is prepared to get involved with a 3D picture composition. The fascination a visualizer has for the work he does is surely the reason why the level of the best technology available today is as high as it is. For many developers and visualizers, spending time with visualization and computer graphics is much more than a regular job.

## Terms

The foundation of the visualization of geo data is the digital terrain model. Further terms for digital terrain models are "digital landscape model", and "digital elevation model".

The different terms include different meanings. A digital terrain model consists of pure geometric information to describe the surface of the terrain. A digital terrain model is also sometimes simply called a terrain model.

A landscape model however can also include information to describe buildings and serves as the basis for analyses of statistical tests. These landscape models (with buildings and vegetation) are also termed digital surface models.

Digital terrain model – pure geometry of the surface of the earth
Digital elevation model – digital terrain model
Digital surface model – terrain model with buildings and vegetation

## Target Groups

It is worth thinking about which target groups one is aiming for with visualization before starting to work on the first draft for a visualization.

What kind of visualization and presentation is made depends on the target group – and the target group comes first in every visualization project.

The choice of the "right" target group can make or break the attempt at good communication.

If we apply the **sender – receiver – model**, we can describe communication as follows:

1. The sender wants to convey a certain subject matter, i.e. he wants to communicate something
2. The channel transmits the information from the sender to the
3. Receiver – who receives the information and decodes it.

The sender can be either the contractor or the executer of a message. Both of these need to communicate their messages in an understandable manner, according to the specific requirements of their positions.

If we continue working with this model, the channel can be represented by the possibilities offered by visualization. These can be digital media as well as film, classical models (model building) or even just a plain sketch. The way in which the information is conveyed is most effective when all parties concerned reach a consensus as to the medium with which this is to be done. This medium should be the one in which a maximum of information is transferred and a minimum of effort is required to create the message and to decode it.

Receivers can roughly be divided into two groups, namely, professionals and ordinary lay persons (of whom the professionals are more challenging from the point of view of the visualizer). Studies have shown that lay persons assess the use of new media in planning presentations as good to very good. The persons taking part in these studies evaluated static visualizations ("pictures") as well as animated sequences ("films") very positively.

A direct comparison between the two forms of presentation clearly shows that animated sequences are favored (on asking which of the two appealed to the viewer more). A remarkable number of those questioned could suggest improvements. One of the main points where there is room for improvement is interactivity. Individual choice of route, of perspective, definition of position and moving in reverse are desired.

The message in the form of an image, which is transported by media channels, must relate to the receiver. Whether the receiver understands a message depends on what he already knows. The more the receiver already knows, the more abstract the message can be, while still allowing the whole image to emerge in the mind of the receiver. Laypersons therefore require more information (i.e. realism) than experts.

**hint Professional and Layperson**

*One can roughly distinguish two kinds of 3 dimensional visualizations of landscape and terrain, namely "technical" visualizations as a means of communication in planning and design among professionals, and "realistic" realizations and representations as a tool to explain/ sell/ in media for the lay person.*

## Fields of Application

Where are digital terrain models used?
Weibel and Heller delineate five main fields of application of terrain models according to occupations[3]:

1. Surveying and photogrammetry
2. Civil engineering and landscape architecture
3. Resource management
4. Geology and further earth sciences, and
5. The military

In the meantime, one can undoubtedly add the games industry as a sixth independent field of application.

While the analysis of the DEM is certainly at the fore in resource management and in the earth sciences, in all other areas, the visualization of virtual terrain is clearly paramount.

Not only the often-described power of images is increasing in importance in digital communication. One can even go so far as to say that one

[3] (WEIBEL, HELLER 1991, Digital Terrain Modeling, Vol. 1. longman Scientific & Technical, Harlow, 1991)

can no longer imagine presentations of complex projects without measures in image creation of some form or another – be it a moving image or a still.

In civil engineering and planning, pictures, or more precisely, 3D visualizations, are mainly employed in the following ways:

In real estate
For competitions
As an aid in discussions and decision making processes in extensive (large) projects before the start of building
In the presentation of projects for public decision making

The faster and better a project can be made clear, the easier it is to represent it and to realize it.
The special abilities and skills of the person responsible in a particular area of expertise also determine whether a project will succeed or not, not only his mastery of a special program or his technical education.

## Why 3D Visualization?

The engineer, architect and planner is not so much simply responsible for the technical realization – he acts as an author, a kind of storyteller or picture author. A discerning contemplation and consultancy should go hand in hand with the planning of a visualization project. Reflecting on the needs of the client and relaying the key aspects of a project swiftly and comprehensibly always have top priority before starting with the realization of a project[5].

One can view the model from different sides
Being able to look at something 3 dimensionally corresponds to our natural way of looking at things and allows, especially in the field of terrain data, a much faster comprehension of complex surfaces and geometry.
The possibility of interaction with 3 dimensional data exists
Problem areas (weak points) of complex constructions are easier to represent 3 dimensionally
Most landscape data already exists 3 dimensionally anyway, and is edited 3 dimensionally in (pure) planning
One can save a lot of time by avoiding misunderstandings
3D visualization enables an "engineer-like" analysis of data

[4] Mach 2000

Construction data can be extracted from the visualized models
One can make things easier to comprehend for non-experts
Concepts and ideas are easier to sell with the help of 3D visualizations.

In civil engineering and planning, many agencies are only contractors and commission specialists to generate visualizations, as the internal operating expense often seems too high.

**hint Similar development to GIS/CAD?**

*It is assumed that the contingent for visualizations in the scope of planning will keep expanding (increasing education in 3D visualization at Universities, cost pressure, etc.). A similar development could be seen a few years ago in the range of CAD/GIS. At first, there was great scepticism about digital media. Today, however, CAD is the norm in almost all sectors of planning.*

In communication by means of 3D visualization, traditional media channels like newspapers and magazines are increasingly supplemented by new media forms. Meanwhile, "new media" have been around so long, that they cannot be considered "new" anymore. These media (www/touch screens) are portals for the presentation of texts, individual pictures, animations and films.

One cannot imagine real estate without the Internet anymore. One look at the real estate section of the newspaper will confirm this.

These media still play a secondary role when informing the public of competitions and other plans, although museums have been proving impressively for years that it is possible to present multimedia-based exhibitions via Internet portals and information terminals in order to awaken the interest of the visitors. Heightened interest and political maturity of citizens will certainly add to the pressure on planners to communicate in an increasingly visual manner. Much as many an expert would like to hide behind his plans, the willingness of decision makers to employ the methods described in the following on visualization is slowly but inexorably growing.

## Consideration of Artistic Concerns

The creation of every image used for communication purposes is subject to the basic principles of an accomplished composition. The clearer these are, the easier it is to convey the matters, which have been translated into an image. In order for this to be so, one has to observe the rules of design, and to keep the question in mind as to who is going to use or view the visualization. To put it another way, one has to consider the role that design and art plays in landscape visualization.

Mike Lin, who has been successfully teaching his workshop "Graphic Workshop for Design Professionals" for many years in the USA (www.beloose.com) gives many suggestions on composition and the use of color in analogue architecture renderings in two books and on his Webpage. Most of these can also be applied to computer graphics, and can be roughly translated for computer graphics as follows:

### Composition

"Start small" – it is best to begin with basic settings for illumination and camera. Getting to know your software by means of small landscape models is a good way to start. One should do without textures and materials at first, and rather familiarize oneself with the geometry. This is easiest when using shades of grey and placing an emphasis on camera positioning and illumination. Once the first obstacles have been overcome, one can start working with larger models and higher levels of resource.

### Less is More

When working on a model for a long time, one risks not only getting lost in details, but also of forgetting what it is all about, namely the visualization of a scheme, an idea or a project. Too many details overburden a scene rather than being of any use. White empty space is not only empty; it helps to highlight important aspects.

The love of details can also inhibit the view of the whole. The art of 3D visualization certainly lies in having the right feeling for how much is necessary to represent the important aspects of a scene.

## Building up a Scene Step-by-Step

The important point is to choose selectively what should be included in the first rendering of an image. Individual and important elements should be singled out and, after leaving sufficient space to the image border, the impact they have should be examined, before starting on the final result.

## Generating Disturbance

Landscape models are predestined to be "disturbed". One should simply avoid clear geometry. The camera should be positioned so that the horizon in the resulting image is broken by "restless" elements. These elements facilitate a natural effect and can help to avoid angular images, especially in computer animation.
This also holds true for transitions of diverse natures. If, for example, a street runs through a landscape, the dividing line between the two areas should rather be interrupted and broken than clear and straight.

## Light and Surfaces

One can get a good feel for a scene if one keeps the materials of a scene simply in shades of grey and one attempt to create day and night situations using only illumination.

## Mass, Weightlessness and Form

Objects are not only perceived according to the way they themselves look; rather, they are perceived in comparison to or in relation to other objects or the surrounding areas of the composition.

This can be the direct comparison of a person in relation to the entrance of a building, or a lawn with trees at its edge.

However, also the weight of an object, the impression of heaviness or lightness, depends on its brightness or luminosity. This can be seen clearly when looking at a mountain (dark and massive) and the bright sky behind (or above) it, which in contrast seems light (weightless). The sky makes it possible for the mountain to seem heavy. Consideration of the interdependencies between light and dark is an important aid in creating an accomplished scene.

## Asymmetry

Asymmetry spices up realistic scenes. Natural scenes are never symmetrical, so symmetry should never dominate in scenes created in 3D. Plants, backgrounds and people can contribute to the creation of "disturbance".

## Hiding the Horizon

It is never possible to see the horizon as a line except when looking at the sea, a large lake or an extremely flat desert landscape. Therefore, one should never show the horizon as a straight line (apart from these exceptions) in artificially created scenes in 3D. If possible, one should cover the horizon or hide it behind other things.

There are bound to be further points, which can be adapted to suit the visualization of terrain and landscape data.

It is advisable to take a closer look at the theory of colors. There are plenty of excellent and recommendable books on the subject. It adds up that there is little sense in dealing with the aspects of the color theory in a black and white book.

# Summary

One can distinguish between two kinds of target groups or receivers according to their level of knowledge. These are respectively experts (professionals) and non-experts (lay persons).

There are diverse fields of application for digital terrain and landscape models. These range from photogrammetrical analysis to the creation of a landscape for a computer game.

It makes sense to create 3D visualizations and can help to avoid mistakes at the start of a planning project, it can make information easily understandable for non-experts and enables the creation of a communication platform for all parties involved in a project.

When working with 3D visualizations, it is important to include artistic aspects in the list of priorities.

An overview and a partly detailed examination of how this can be done can be found on the following pages.

# Fundamentals and Data Source

*A short overview of the historical development of landscape visualization, and a few facts about the sources of data.*

## The Development of Landscape Visualization

One of the oldest cartographical depictions of a landscape was discovered in 1963 during excavations in Catal Höyük, in Central Anatolya. The mural painting is 3 meters long and dates around 6200 B.C.

Experts presume that the map represents the city of Catal Höyük. Apart from houses and terraces, the twin peaks of the volcano Hasan Dağ are also depicted.

**Fig. 2.** Catal Höyük (mural painting) – one of the first cartographical representations

In cartographical research, these "molehills" are regarded as the first representations of terrain (E. Imhof: Cartographic Relief Presentation).

Mountains in the form of 'Molehills'.

A symbolized mountain range. The symbols are left, right, downhill and uphill orientated

Melt down of 'hill symbols' to an over-all form

Perpendicular hills

Squamous aggregation of hill symbols

shading and variations of mountain and hill form, Germany, 16. century

**Fig. 3.** Hill and mountain shapes and their development over the centuries [1]

The first forms of representations of terrain where profile-oriented representations of reality. This kind of profile-based representation of terrain was adhered to for many centuries. During this time, maps resembled two-dimensional pictures with cities, castles, monasteries, forests and mountains.

[1] Empathized from Eduard Imhof, 1982, sketch by Yves Maurer

## hint Landscape Painting and Cartography

*It is assumed that landscape painting, an important part of painting, and cartography share the same origins.*

An innovation typical of Renaissance art was the use of perspective in the depiction of landscapes. It also leads to more realistic, 3-dimensional representations of shapes in landscapes. The maps of Tuscany, which Leonardo da Vinci painted between 1502 and 1503, are good examples of this development. This was the first time that terrain forms were represented from a bird's eye perspective.

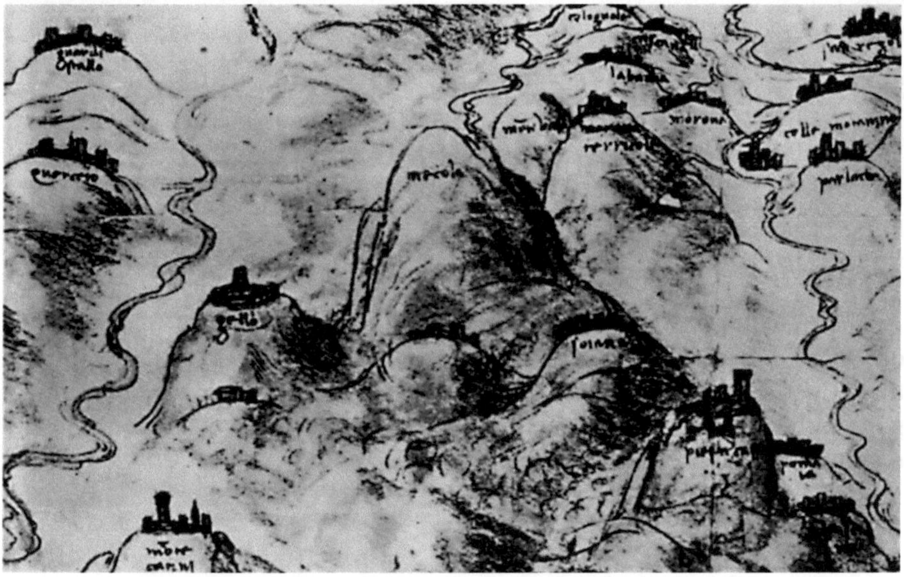

**Fig. 4.** Extract from Leonardo da Vinci's maps of Tuscany [2]

A further pioneer in the representation of landscapes and gardens was certainly the artist and master of landscape gardening, Salomon de Caus, who unfortunately never completed the famous garden "Hortus Palatinus" of Heidelberg Castle, Germany.

[2] by courtesy of „Digitalisierungswerkstatt- Universitätsbibliothek Heidelberg“

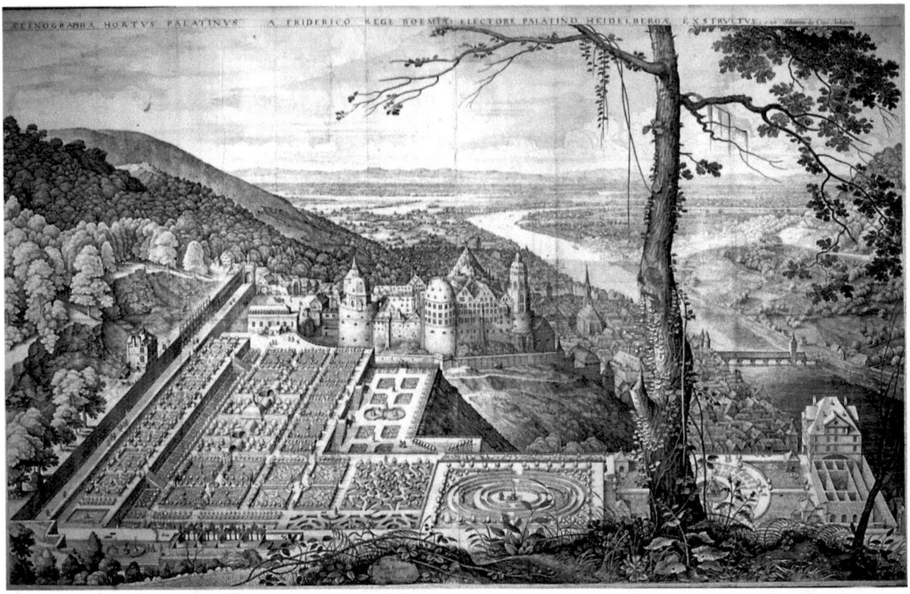

**Fig. 5.** Copperplate engraving of the Hortus Palatinus of the Heidelberg Castle gardens, 1620[3]

The transition to a completely top-view representation of maps – without any side views, only took place shortly before the end of the 19th century. Ever since, diverse kinds of hatching have been used to represent topography (coloring of height layers, shading etc.), which facilitate the reading and comprehension of maps considerably.

[3] Source: Uncolored copperplate engraving of M. Merian, 1620, - by courtesy of „Digitalisierungswerkstatt- Universitätsbibliothek Heidelberg"

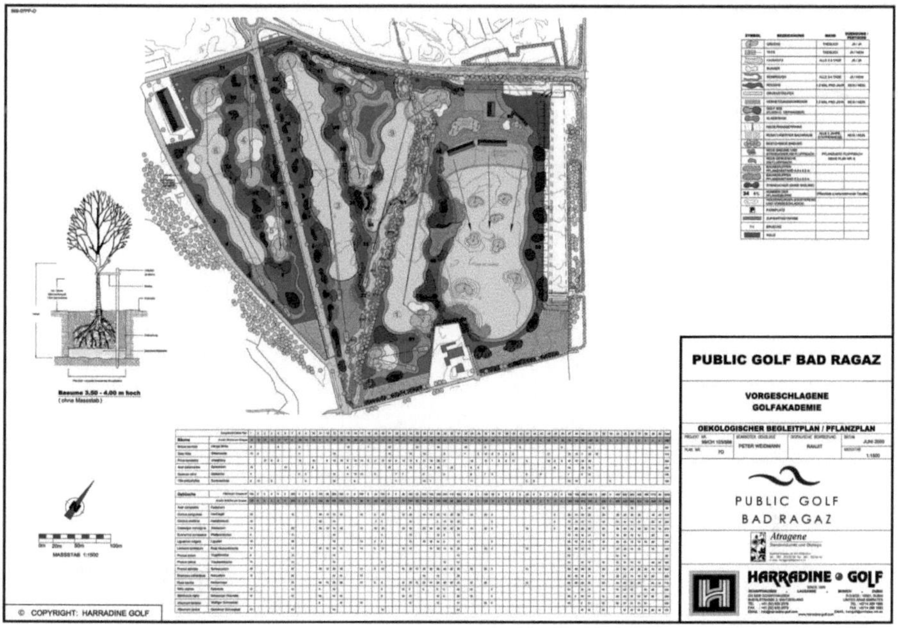


**Fig. 6**. Modern planting plan with laminar color enclosures and shadings

Parallel to this already very clear form of presentation, contour lines were developed as a possibility of representing information about the height of a terrain. Contour lines are lines which connect points of the same height above a certain geodetic point (sea level), in order to represent the relief of terrain. The advantage of this method is that it conveys quantitative information about the relief of an area.

Pieter Bruinss was the first to use contour lines, in 1584, in his sea chart in the Netherlands. Nowadays, they are not only used in sea charts, but also in recreational areas (hiking and ski tour maps) Contour lines are regarded as the cartographical form of representation of terrain.

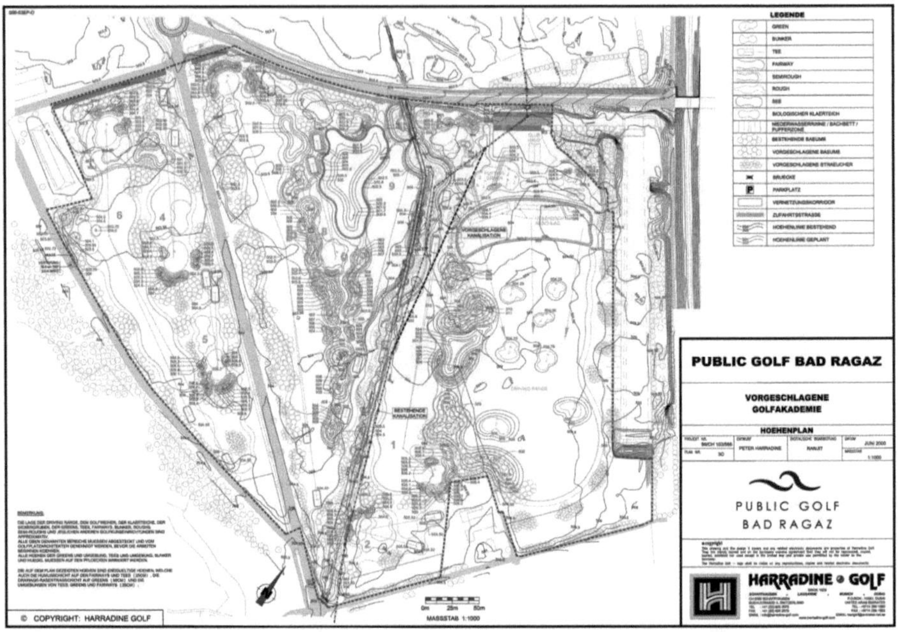


**Fig. 7.** Modern chart with contour lines of a golf course near Bad Ragaz

Because contour lines convey a graphic impression of the shape, gradient (incline) and elevation of an area, they are also used in landscape architecture and civil engineering in the representation of a planned terrain to be graded.

Grading Regional planning targets are:

Leveling off of surfaces for houses, parking lots, soccer fields etc.
Development of accessible points on different levels (streets, roads ramps with maximally 12% respectively 6% inclines
Creation of special situations (e.g. golf courses) or solution of technical problems (slopes in road and railway construction).

Contour lines can be constructed by interpolation between a series of points with known heights. The determination of height points is carried out with the help of a surveyor's staff, an automatic level and a geodetic point. Horizontal or vertical angles are measured by means of a theodolite.

The famous landscape gardener Humphrey Repton pursued the canvassing of clients in 1788 already, using business cards about landscape design.

The flyer, of which Repton had over 1000 printed, shows him standing in the foreground in front of a theodolite, giving directions to his workers digging on a construction site.

*„Completely engraved, it shows an elegant Repton with a theodolite, directing labourers within an ideal landscape that is derived from Milton's L'Allegro." (G. Carter, P. Goode, K. Laurie: Humphry Repton Landscape Gardener 1752-1818. Sainsbury Centre for Visual Arts Publication, 1982, P. 12/13).*

Humphrey Repton is, in addition to Lancelot Brown and William Kent, one of the most influential representatives of the English landscape gardening style (1720 – 1820), which also became fashionable on the European continent after 1750. During this time, great landscape parks such as Stowe in England or Woerlitzer Park in Germany were created.

**Fig. 8**. Flyer of the Red Book by H. Repton

Projects always mean great changes in landscape. New lakes and hills with houses and bridges embedded in them, accentuated by elaborate planting, had to be advertised and sold to the contractors. In this endeavor, the methods of landscape painting were used more and more frequently.

Repton left behind numerous records of his projects, which became famous under the name "Red Books".

Fig. 9. Cover of the Red Book by H. Repton

Theses books are bound in exquisite red Moroccan leather, and they depict his plans verbally and visually. On the first pages of any one of the Red Books, Repton describes the particular project elaborately in writing. Usually the text is followed by watercolor drawings of the existing situation. The special thing about the drawings is that a part of them can be removed by means of a page which lies on top. The situation which has been planned and suggested by the landscape architect lies beneath this removable page.

Repton always presented his projects personally to the client by showing his Red Books. Usually the client was so impressed by the clear presentation that he gave him the commission immediately. The Red Books remained in possession of the client and were charged to their account. Thanks to his innovative marketing strategy based on a 3D visualization technique, Repton, who lived from 1752 till 1818, could build numerous large parks in Great Britain, and so he became one of the most successful landscape architects of his time.

**Fig. 10.** Extract from the Red Book. By means of variations which were placed on top, plans for a project were impressively shown to the client.[4].

The first digital elevation models (DEM) in the field of cartography were created by Miller and Laflamme in the Photogrammetry Laboratory of the Civil Engineering Department of M.I.T. in the late fifties in the course of research (see Miller C L, Laflamme R A (1958) The Digital Terrain Model – Theory and Application. Photogrammetric Engineering).

The Massachusetts Department of Public Works supported the project financially in collaboration with the Bureau of Public Roads, with the goal in mind to employ terrain models in road construction.

## Basic Data

The necessity to visualize landscape data has certainly always been especially challenging. The most laborious part of it is however the acquisition

[4] Source: Red Book, Schweizerische Stiftung für Landschaftsarchitektur SLA, HSR Rapperswil

of the basic data. Basic data in this context refers to measuring data, and not to data connected to planning.

As exciting as it may be to create an arbitrary landscape according to one's own wishes – in whichever form this may be done – planners usually have to edit, convert and process reality-based data first.

The greatest challenge in most areas is certainly reality-based data. This data is not always clearly defined, it usually comes in large packages – large packages meaning in this context ASCII files > 100 Mbytes, and not everyone has at their disposal the necessary tools to view and process this data. This issue will be dealt with in detail later on.

## Geometrical Data

Geometrical data is usually the foreground of a project. Geometrical data, which is used to represent geographical information, consists of points, lines and polygons.

As a basic principle, there are three types of geometrical geo data of this kind:

**Vector data:** Contour lines (isolines) or polygons created from multipoint sampling, which find their way into the model by forming edges, for example embankments, streets, walls etc. significantly determine the contours of a landscape

**Grid data:** Picture information in the form of BMP, JPEG, GIF, georeferenced TIF etc. usually provide height information, aerial views or further information on the appearance of landscapes.

**Information of Matter:** Annotations or additional information transcending a purely topographical representation. This information about matter or attributes includes information from statistical analyses for example.

### hint Grid data

*The concept "Grid data" can lead to irritations, as it refers to different things, which are similar to each other. On the one hand, grid data refers to picture information or pixel images (JPEG, GIF, TIF). On the other hand (in the field of terrain models), a grid refers to a dot matrix which is used to define a terrain model. Picture elements (pixels) of a pixel file are described by their color values, such as RGB. The elements of a grid for terrain models are furnished with height information. So don't let yourselves get unsettled!*

## Aerial Views, Satellite Images

Aerial images and satellite photos constitute the basis for the creation of digital terrain models on the one hand; on the other hand, they are an important part of 3d visualizations, as they can be used as material and thereby enable the establishment of a relation to reality without too much sophisticated work in modeling.

### *Aerial views*

In remote sensing, pictures are taken of existing terrains from a bird's eye point of view. Usually, these are taken during flights over a certain terrain.

These flights are usually undertaken either in order to selectively photograph a single object, or to systematically cover a specific area. In the latter, the area is flown over in parallel strips which must overlap each other (30 – 50%). These overlapping strips are then assembled and analyzed.

The analysis is usually carried out using one image, or stereoscopically. Orthophotos should be used in stereoscopy. Here, height information is derived from two images with different points of view.

**Fig. 11.** Example of an aerial image

If there are no orthophotos at one's disposal, "normal" photos have to be equalized. This means that the distortion of a photo which arises from the curvature of the lens is corrected by appropriate measures (nowadays normally using software).

In Germany, the Bundesamt für Eich- und Vermessungswesen creates digital terrain models using aerial photo information. Using stereo photographs, points in a defined 3D coordinate system are calculated and a grid DTM is interpolated from these.

## *Satellite Photos*

A satellite photo is a small scale image of the surface of the earth. Two kinds are distinguishable:

Fig. 12. Example of a satellite photo

1. Passive Systems (photographic cameras, digital or CCD cameras, infrared and multi-spectral sensors): They register sunlight or other forms of radiation reflected from the earth in the form of geometrical images.
2. Active Systems: They radiate self-generated electromagnetic waves (radar and other radio waves, laser) and scan their diffuse reflection[5].

The production and handling of satellite images is a standard method in remote sensing and many geo sciences. It also helps in the cost-effective production of maps and the analyses of wide areas in environment protection. The classification and interpretation of satellite image information provides important databases of geo information systems (GIS).

## Laser Scanner Procedure as Basis for Data

In order to obtain information in 3D of large areas right away, flying over the area concerned is done with laser scanners. Sensors scan the topography of a surface directly and deliver 3D information ad hoc in the form of scatter plots or point clouds. There are diverse scanner types on the market.

These are continuous wave Lasers and Pulsed Lasers. The term continuous wave refers to a laser which produces a continuous output beam.

Pulsed lasers measure the first and last reflection of the signal separate from each other. The high level of penetration makes it possible to measure the forest floor as well as the crowns of trees separately.
The flying altitude averages 1000 – 1800 m. This enables an exactness of 0,1 – 0,3m.

[5] www.wikipedia.org

The result is a grid model. In combination with orthophotos, one has a good basis for the construction of a digital terrain model.

**Fig. 13.** Diagram of a laser scan flight

## GPS as a Source of Data for Digital Elevation Models

GPS[6] is the abbreviation of NAVSTAR GPS, which in its turn stands for "NAVigation System with Time and Ranging Global Positioning System". GPS presents a solution to one of the oldest and most profound problems of humanity. It answers the question:

"Where in the world am I?"

One could imagine that this question is easy to answer. It is relatively easy to spot one's own standpoint in relation to the surrounding objects by

[6] GPS was developed 1978 by the American army (Glonass is the Russian counterpart to GPS, the European system is called Galileo)

looking around. But what happens when there are few surrounding objects?

What happens in the middle of the desert or at sea? For many centuries, this problem was solved by means of navigation according to the sun and the stars. These methods work well, but they also have their limitations. When clouds cover the sky, for example, the sun and stars cannot be seen and orientation is not possible.

GPS is a satellite-based system which provides the user with an exact location (longitude, latitude, and altitude) in any weather, day or night, anywhere on Earth using a constellation of 24 satellites.

GPS has numerous advantages compared to traditional surveying methods:

1. Intervisibility between the points is not necessary
2. Availability at any time of day or night and in any weather
3. Decidedly more work can be done in less time by less people
4. GPS provides results with a high level of geodetic exactness

With GPS, very accurate measurements of elevation can be attained.

**hint About Accuracy**

*This is an appropriate place to examine the term "accurate" more closely. For a wanderer in the desert, "accurate" means about 15m tolerance, for a ship in coastal regions, accuracy means a tolerance of 5m and for a land surveyor "accurate" is 1cm or less.*

GPS can be used to attain all these kinds of accuracies in all these areas. The difference lies in the kind of GPS receiver and the technology used. Nowadays, there are products on the market which have been developed for users with little or no knowledge of surveying. With the use of an antenna, one can measure elevations with them. The scatter plot measured can be transferred to the computer via Bluetooth, and a terrain model can easily be created with this information. The accuracy of the elevations depends on the technique used.

GPS measurements can be done in different grades of quality, which can be distinguished according to the techniques used. These are:

Simple navigation
Differential GPS (DGPS)
Real-Time GPS (RTK)

### Simple Navigation

This is the most simple technique used by GPS receivers to provide the user instantaneously with a position and height. The accuracy achieved lies under 100m for a civilian user (usually around 30 – 50m). Receivers which are usually used for this operational mode are small, easily transportable units which can be hand-held and cost little.

An accuracy of 1 – 2m can be attained with the use of an antenna. However, for most terrain models this data is too inaccurate.

### Differential GPS (DGPS)

Differential GPS is an enhancement to the GPS and means that a second stationary receiver is used in order to correct (control) the measurements made by the first. The level of accuracy which can be achieved in this way increases up to one meter. However, it is important to remember that the further the receivers are apart from each other, the higher the level of inaccuracy is.

### RTK Real-Time GPS

RTK stands for "Real Time Kinematic" and is a kinematical (not static) "on-the-fly" measurement which takes place in real-time. A stationary "base station" and a mobile "rover" are employed when using this method. The base station has a radio link and it sends the data which it receives from the satellites. The rover has a radio link too, which enables it to receive the signal from the base station. Furthermore, the rover receives data directly from the satellites via its own GPS antenna. These two sets of data can be processed in the receiver of the rover in real time, i.e. immediately in order to resolve any ambiguities and in doing so to achieve a highly accurate position in relation to the base station. Find out more about RTK under the Thales website.[7]

### Real Time Error Detection and Correction

DGPS devices are able to receive two GPS frequencies (L1 + L2). The difference between the signals gives the rover information about what is going on in the atmosphere. The position can hereby be corrected in real time.

These devices are much more expensive than L1 devices. As an alternative to L2 devices, it is worth considering the use of L1 devices which are

[7] http://products.thalesnavigation.com/

able to receive correction data, thereby enabling them to heighten the accuracy of the position.[8].

Correction data comes from a base station and can be received via mobile. A base station, the position of which is exactly determined, constantly redetermines its position via GPS. A correction factor can be evaluated using the difference between the absolute and the newly determined position. This can be received via the GSM network (mobile telephone network) in real time. The signal can be transmitted to the rover via mobile and Bluetooth.

### hint Correction using EGNOS

*The European alternative Galileo includes a system called EGNOS, which transmits correction data which has been checked for feasibility. This data can be interpreted and processed by a large variety of devices.* [9].

### *Post Processing Correction*

Collecting data and correcting it on the following day is another possibility. For post-processing via the internet, correction data is only available 24 hours after data has been collected, by means of which one can correct the data which has been collected the day before. For terrain models this is alright as the accuracy is acceptable. Furthermore, the terrain model usually does not have to be finished within 24 hours either.

### hint Incidentally: GPS

*Originally GPS was intended to aid military missions at all times and in any place on earth. However, it soon became clear that GPS could also be useful to civilians – not only in helping them to determine their own positions. It turns out that the first two main areas of application by civilians were in navigation at sea, and in surveying. Nowadays, GPS is used in such diverse areas as autonomous vehicle locating and navigating systems, in the field of logistics in transportation businesses ("fleet management"), and the automation and controlling of construction site equipment.*

---

[8] Wikipedia.org: The typical nominal accuracy for these dual-frequency systems is 1 centimeter ± 2 parts-per-million (ppm) horizontally and 2 centimeters ± 2 ppm vertically.

[9] http://www.esa.int

## Data Evaluation

Having collected all necessary data and being able to read these does not mean that all obstacles in creating a landscape or terrain model have been overcome. Often, measures have to be taken on this data before work on the final visualization can begin.

An excellent way of gaining a quick overview of the data at hand is the use of a GIS. GIS stands for Geographic Information System, and products such as ESRI's ArcGIS or Civil3D (Map3D) by AutoDesk are probably known to most of our readers.

### Tip Open Source

*The examples in this book were created using the above programs, but it is definitely worth taking a look at what is available in the field of open source.*[10].

### GIS Tools

Naturally, there are different tools which can be used when handling geo data. For some, a small CAD-based terrain modeler may suffice, others may need a personal geo database to be able to manage their projects. One thing is certain: GIS is of prime importance when dealing with geo data.

GIS is a tool which cannot be avoided by specialists whose work involves anything to do with geo data. It should be the norm in the same way that the use of word processing software is.

There is nothing better than a GIS for the first evaluation of geo data, as a GIS can utilize just about any forms of data which are relevant to 3D modeling. These include:

Geo data – vector-based geo data, geo databases and most current geo data formats
CAD data e.g. DWG, DXF, DGN
Facts in any kind of database
Open surveying formats, usually in the form of ASCII files

[10] Have a look at http://opensourcegis.org/ or http://grass.itc.it/

## Breaking lines

If a DTM that is provided for visualization is not "complete", and the visualizer still has modifications to make, it can be helpful to address a couple of aspects of breaking lines.

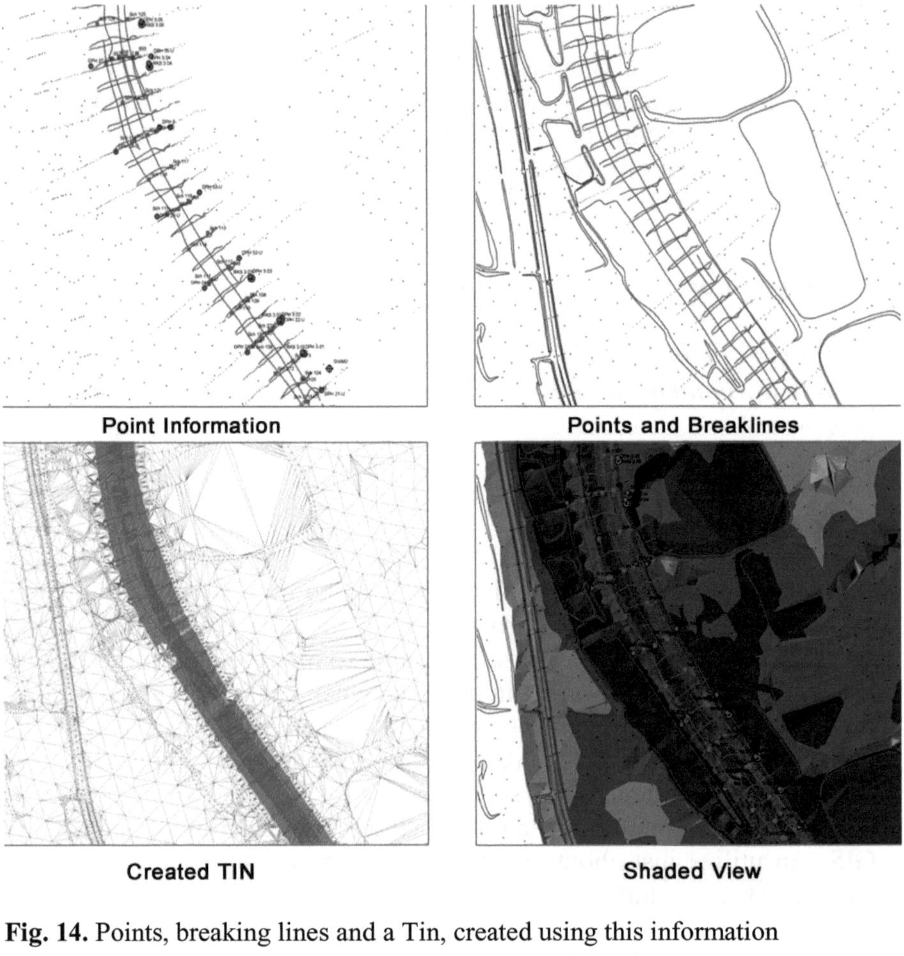


**Fig. 14.** Points, breaking lines and a Tin, created using this information

Breaking lines are traverses which describe parts of the terrain accurately, such as the course of streets or dams for example.

If, for example, a grid DTM forms the basis, breaking lines cannot be depicted, which can lead to inaccurate representations.

In a TIN, which is not bound to a defined grid, any line information can be accurately intermeshed and represented without any difficulties.

## Coordinate Systems

The surface of the earth, which is approximately the shape of a sphere, must be depicted on a flat surface. To find any place on earth at any time one has to define a coordinate system which describes ones position easily. The most commonly used coordinate system is the latitude, longitude, and height system. In this system, the Prime Meridian and the Equator are the reference planes used to define latitude and longitude.

Latitude (Lat.) is the angle between any point and the equator. Lines of constant latitude are called parallels. They trace circles on the surface of Earth, but the only parallel that is the large circle is the equator (latitude=0 degrees), with each pole being 90 degrees (north pole 90° N; south pole 90° S).

Longitude (Long.) is the angle east or west of an arbitrary point on Earth: The Royal Observatory, Greenwich (UK) is the international zero-longitude point (longitude=0 degrees). The anti-meridian of Greenwich is both 180°W and 180°E. Lines of constant longitude are called meridians. The meridian passing through Greenwich is the Prime Meridian. Unlike parallels, all meridians are halves of great circles, and meridians are not parallel: they intersect at the north and south poles.

By combining these two angles, the horizontal position of any location on earth can be specified.

For example, Baltimore, Maryland (in the USA) has a latitude of 39.3° North, and a longitude of 76.6° West (39.3° N 76.6° W). So, a vector drawn from the center of the earth to a point 39.3° north of the equator and 76.6° west of Greenwich will pass through Baltimore.

Another very well known coordinate system is the UTM projection system.

*The Universal Transverse Mercator (UTM) projection system is a grid-based method of specifying locations on the surface of the earth. It is used to identify locations on the earth, but differs from the traditional method of latitude and longitude in several respects. The UTM system is not a map projection, but rather employs a series of zones based on specifically defined Transverse Mercator projections (Wikipedia.org).*

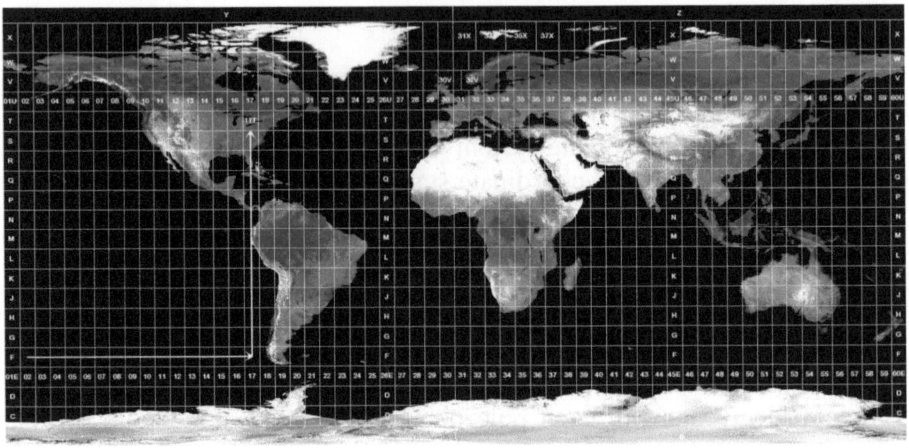

**Fig. 15.** UTM -This image was constructed from a public domain Visible Earth product of the Earth Observatory office of the United States government space agency NASA[11].

The origin, where x and y equal zero, of the Swiss national coordinate system, is located in the vicinity of Bordeaux in France. All coordinates of places in Switzerland are therefore positive. Because all points are located in the first quadrant, all signs can be omitted. Furthermore, it is impossible to mistake a y value for an x value, as the y values are always larger than 480 km, and the x values are always smaller than 300 km.

The advantage of working with the national coordinate system is that the data (e.g. planned positions of trees) is defined in CAD using national coordinates, and can be passed on directly to the geometer to be staked out accurately on the construction site.

[11] http://en.wikipedia.org/wiki/Image:Utm-zones.jpg

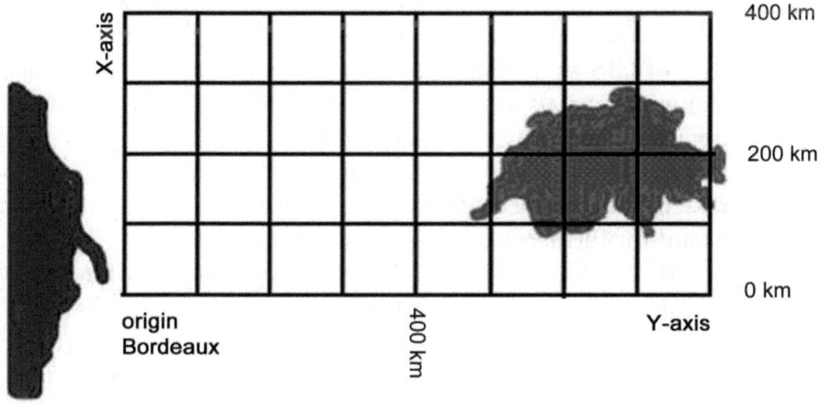


**Fig. 16.** The Swiss Local Coordinate System

### hint Easting and Northing

*The term easting is a land surveying term, referring to Cartesian distance measured eastward, or 'X' value. This comes from a methodology for computing coordinates and areas, known as the Method of Latitudes and Departures. The counterpart is northing, representing the 'Yy' value. Both values are given in consistent linear units of measure, such as meters or U.S. Survey Feet. Typically, in mathematics, and in land surveying in most parts of the world, an XYZ convention is used for representing locations in plane surveying, however in the United States for land surveying applications, an NEZ convention is also quite common (Wikipedia.org).*

## Interfaces to 3D Visualization

The difficulty in realizing visualization projects using currently prevalent terrain modeling tools lies in the quality of the result. Already GIS applications can generate excellent 3D depictions and evaluations of data.

Visualizations based on GIS are usually the form of exchange among experts. These programs are however neither made for, nor are they especially suitable for high quality visualizations.

Particular programs are specialized to work with geo data, such as the World Builder by Digital Element (www.digi-element.com) or diverse plug-in providers for all current 3D programs. However, the conversion of data often fails due to problems with readability of public data such as Atkis (or similar forms), which were not conceived for visualization. Or to be more precise: for which no utilizable interfaces for visualization exist.

Here, the use of GIS as a "data converter" helps. Furnished with the appropriate functionality, it is usually easier to read geo data and to work with and export this data for use in visualizations.

Many planners create their digital terrain models using specialized tools which have been programmed for CAD platforms. These digital terrain models are meant for further use in 3D applications.

Such programs, e.g. Civil 3D in an AutoCad application, offer a respective output of data, either as DWG or DXF data, or they offer direct interfaces to the world of 3D visualization.

### hint Civil 3D

*Civil3D by AutoDesk is already provided with the complete VIZ renderer and spares one the conversion to visualization in specialized programs.*

### *Data Transfer*

Usually, the 3D user or visualizer does not have a GIS application. Nonetheless, there are ways of handling such data.

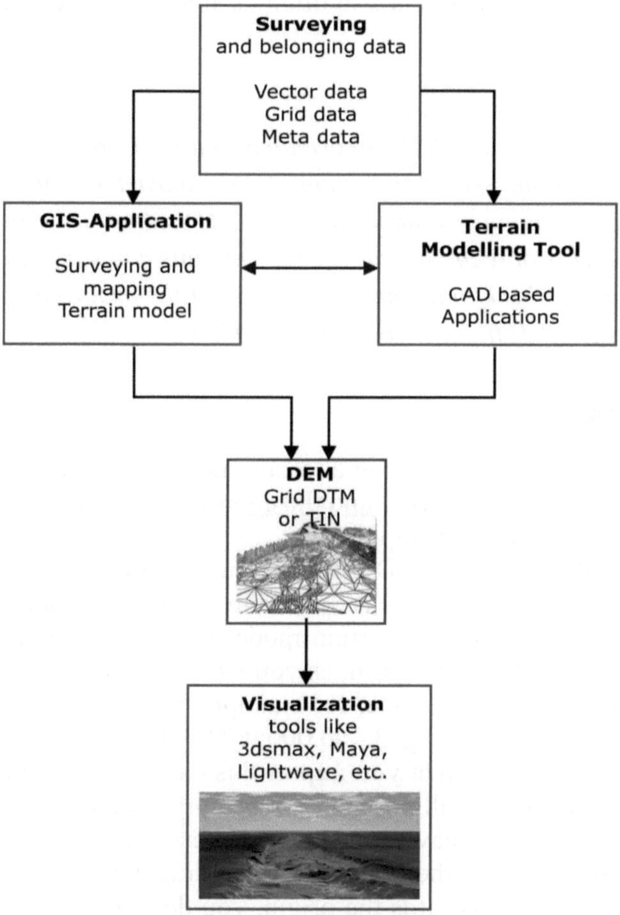


**Fig. 17.** Data flow and Work flow for generating 3D visualizations based on Geo data

The first approach leads to a discussion with the client about a suitable exchange format. Current GIS products make it possible to convert a grid DTM into TIN and export CAD data like DWG or DXF.

What also (almost) always works is the output of terrain information as nodes or mass points (or vertex[12] in XYZ) in the respective coordinate system (e.g. WGS84, World Geodetic System).

Most 3D programs are able to import CAD data, at least DXF (Data Exchange Format by AutoDesk) data or ASCII point data. VRML (Virtual Reality Modeling Language) is also often used to exchange data.

**hint ASCII?**

*The American standard ASCII (American Standard Code for Information Interchange) encompasses a maximum of 256 characters and can be read by (almost) any computer system. The AutoCAD format DXF is formatted in ASCII. ASCII is only the language; the contents can vary of course. Here, you should check in the documentation of the program you are using which kinds of ASCII files can be read or written.*

### *Coordinates*

Care must be taken when creating a terrain model, as relative or user coordinate systems are usually used, and the coordinates inflate the memory.

Let's take a look at an example:

You are working with a CAD program and WGS coordinates in order to describe your landmarks. You have a terrain model and would like to visualize it in a 3D program. The terrain model is furnished with the original coordinates in your CAD program, as you would like to evaluate construction information. Let us imagine, for example, that the size of the coordinates is $10^6$ units (meters), e.g. 3.400.000, 5.500.000.

Let us further imagine that you import this data into a 3D program. You will find that the values will possibly not be correct anymore, as the values with which the polygons have to be built are very high.

If, however, you shift the objects in your drawing before exporting by e.g. $10^6$ units (meters) towards the origin, you decrease the values without changing the depiction itself. This means, however, that you can only determine co-ordinates in your 3D program by using this shift vector. Furthermore, standard problems such as "inverted normal orientation" and associated errors in material assignment keep coming up.

[12] In 3D computer graphics, a vertex is a point in 3D space with a particular location, usually given in terms of its x, y, and z coordinates

### 3D Representation

Usually, a 3-dimensional visualization of landscape and geo data is based on a digital terrain model. Additional terms for digital terrain models are "digital landscape model" or "digital elevation model". These different terms also have different data contents. A digital terrain model consists of purely geometrical information describing the upper surface of the terrain, whereas a landscape model can also include information describing buildings, and often forms the basis for the evaluation of statistical analyses.

As a basic principle, digital terrain models are distinguished according to the way in which geometrical data is managed, and according to two important criteria:

Grid DTM
TIN

### Grid DTM

The term grid DTM refers to an even grid with a defined "width" (e.g. 25 m, 50 m etc.). Each point on the grid is allocated a height. The advantage of a grid DTM is that the construction of the 3D data can occur by means of a matrix very quickly, and that the representation does not require too much memory space. This is possible because one does not need all coordinates for every point.

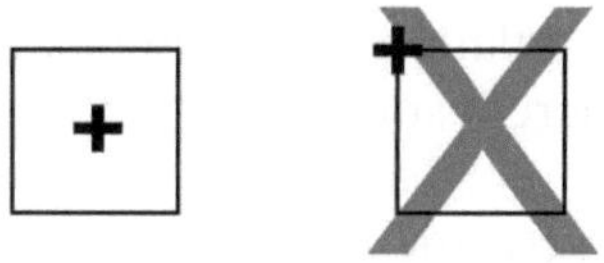

**Fig. 18.** Point of reference of grid elements

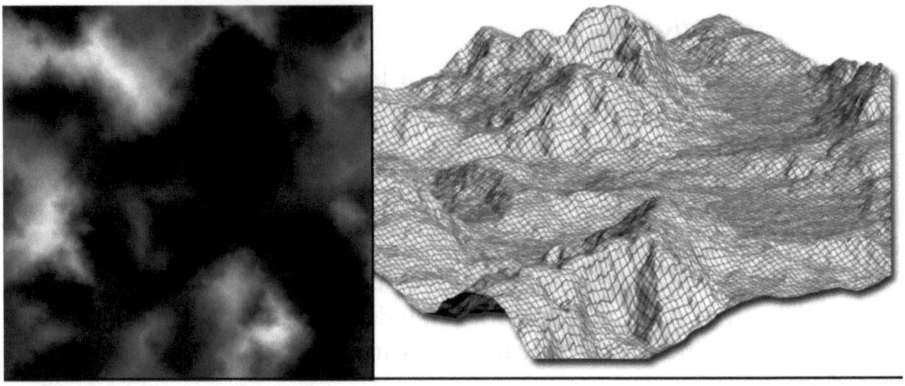


**Fig. 19.** DTM 8 Bit grey scale picture (even pixel grid with attribute values for the terrain elevation/height) and the resulting wire frame model

## *TIN*

The Triangular Irregular Network (TIN) is used in the representation of digital terrain surfaces.

Further terms are "Digital Elevation Model" (DEM) or "Digital Terrain Model" (DTM).

Take note: a TIN is always the basis for all of these models!

The structure of TIN data is based on two elements: points with X, Y and Z values and edges. By means of triangulation of the points, a mosaic consisting of triangles is created, and a coherent surface is produced.

The more points a model has, the more possible ways there are of creating triangles. Inaccuracies are the result of long, narrow triangles, and therefore these should be avoided. This is why one needs a triangulation algorithm which is able to create adequate triangles. Apart from avoiding long, narrow triangles, an important criterion for the algorithm is that a maximum amount of triangles whose sides do not intersect are created. The Delaunay triangulation meets these needs best. If the program, e.g. Civil 3D by Autodesk, generates a triangulated DTM on the basis of point data, the DEM is calculated by means of the Delaunay triangulation of the points. With the Delaunay triangulation, no other point exists within the circle defined by the angular points of any given triangle. Delaunay triangulation has many advantages compared to other methods, namely:

The angles of the triangles are relatively similar, which means that long, narrow triangles and the potential numerical inaccuracy connected with these are avoided
It is ensured that every surface point is as close as possible to a knot
Triangulation is independent of the order in which the points are calculated

TINs are created using points, lines and surfaces. They are largely generated automatically. The origin of the data can vary widely (attained using GPS oneself, or digitalized using a map, or obtained from the land surveying office).

A TIN surface is always made up of connected triangular surfaces. The height of every point on the surface can be interpolated from the x, y and z values of the surrounding triangle. This also means that the slope and the contents of the surface can be determined.

It is very important to note that the accuracy of the model and calculations depends on the original data.

One of the problems when generating a TIN automatically based on contour lines is that sometimes horizontal (flat) triangles are created. Triangles should however be made from points with differing heights. As there are often no suitable points close enough on the neighboring contour line, the nearest point on the same contour line is used. This results in a flat surface, where in reality a slope is to be found. This error leads to incorrect results in the evaluation. Usually, DEM programs have commands with which triangles can be turned. A visual check or knowledge of the terrain in reality is necessary in spite of the possibility of generating a TIN automatically.

TIN also offers the possibility of representing a terrain in diverse grids. It still has the advantage, as opposed to grid digital terrain models, among other things, that breaking lines can be integrated. Breaking lines represent edges in the model.

Retaining walls, which prevent landslides in steep terrain, are classical breaking lines. Furthermore, the upper and lower edges of embankments are defined by breaking lines. Breaking line data affects the triangulation of the DTM. The program will connect points to form one side of a triangle, even if the points have a breaking line between them, although this violates the character of Delaunay triangulation. Retaining walls have more than one Z value for one X and Y point. One Z value represents the upper edge of the wall, and the other represents the lower edge of the wall. Due to the fact that only one Z value per point can be recorded when triangulating, the best way to by-pass this problem is by using two lines which are very close to each other, of which the first line represents the lower edge, and the second line represents the upper edge. As it is only possible for one z value to exist for a point defined by x and y in a DTM, the model is in actual fact 2.5D, and not 3D. Because the calculation occurs within the plane, and the real location of the data is not allowed for, constructing tunnels or protrusions using normal DEM programs is very difficult and only possible when using tricks.

## Summary

This chapter gives a short overview of the development of landscape visualization, and a first impression of the basic principles when dealing with data.

Points, lines and polygons form the geometrical basis of a digital terrain model.

3D models can be generated on the basis of aerial photographs and satellite images (e.g. stereoscopy). At the same time, they are used as textures for the final 3D models.

Geo data can be collected in diverse ways. The classical surveying method is one way; the use of GPS and the generation of scatter plots by means of laser scanners are two further ways of collecting geo data.

Geo data must be prepared for visualization before it can be used in a DTM. Diverse tools are used to do this. Furthermore, the underlying relative or user coordinate systems of the respective data sets have to be taken into account.

Data transfer to visualization tools is usually carried out via CAD formats, such as DWG or DXF or other suitable importers.

For visualization, the most commonly used kinds of DTM are grid DTMs and TINs.

# 3D Visualization of Terrain Data

*This chapter focuses on five main topics. These are the import of existing geo-data, the creative modeling, some information concerning materials, a few words about distortions and their animation.*

## Data Import of a DTM

Only very rarely will the 3D modeler be able to create a terrain model as he likes. In most cases he is confronted with the problem of how to represent existing data or how to use this data as a basis of a planned project.

If, as mentioned before, a GIS is at his disposal, normally the data export for further 3D-visualization is no problem. If there is no GIS available, he has no choice but to fall back on current formats of 3D visualization programs.

In this field a lot of progress has been made. Most 3D programs are able to import the following data directly:

CAD interface – mostly DWG and DXF
Points – mostly ENZ[1] or raster data[2] as ASCII file
VRML interface – data format for the description of interactive 3D-worlds
Various 3D formats – interfaces between 3D tools, - the most common being OBJ[3], 3 DS, and LWO
Import as a RAW[4] file or as a grayscale picture.

---

[1] Easting Northing Z (Elevation) is a column ASCII format)

[2] DEM as header defined ASCII file

[3] OBJ – Alias Wavefront data format, 3DS – 3D studio DOS format – however, due to its constraints/ limitations, the 3ds format is only suitable for small terrain models. LWO – Lightwave Object Format.

[4] RAW – is a raw data format which, depending on the make/manufacturer, compresses without loss and is used in digital cameras. It supports CMYK-, RGB- and grayscale pictures with alpha channels as well as multi-channel- and lab-pictures without alpha-channels.

**Tip Consulting the client**

*Before starting with the actual work, it is worth while getting together with the client. Discussing how best to convert the plans, which data is required and the pros and cons for visualization can facilitate the visualization process tremendously. It also improves understanding by preventing a recurring question : "But you can do that by just pushing a button can't you?" Geo-data is often provided by the customer. It is advisable to define in advance which exchange formats you can best work with, in the course of a short talk, preferably in writing. You will be surprised at how much can be achieved this way.*

Nearly all terrain modelers and GIS-tools are able to generate CAD - data. If this is not the case, there is always the alternative of a data exchange via VRML-format (VIRTUAL REALITY MODELING LANGUAGE). In the unlikely case that even this option is not at your disposal, almost certainly you can generate a triple data set (XYZ). However, one has to keep in mind that when importing terrain-data via a triangulation existing breaking lines may get lost and that there is a risk of errors appearing in the newly created terrain model.

**Tip Data exchange in 3ds max**

*In 3ds max it is not possible to "save under previous version". Here a little script by Borislav Petrov "Bobo" called "Back from five" can be helpful. This little program enables you to save all 3ds max data in a executable script. Download and info can be found under http://www.scriptspot.com/.*

## Import of an existing DTM as a triangulated TIN

An often-encountered request in real life can be described as follows:

The terrain model has already been drawn up. The data is provided on request. An important prerequisite is the representation of all important breaking lines in the visualization of the model. The visualization serves as helper for decision for a public discussion.

The data base has been compiled in a CAD-based terrain modeler. This means that the terrain data is given as a CAD data set. Generally the CAD data includes a coherent mesh and is given in DWG or DXF format. Most

PC CAD programs are able to read and write these data formats of Auto-Desk.

Before importing the data, it is worthwhile to test it with a CAD application. It would be optimal if the customer could do the necessary processing beforehand.

If this has not been done yet, you have no choice but to test the data yourself before further processing. The criteria for a successful import can be put down as follows:

Is all accessory data of the DTM in one layer? If not, it is recommended to put these on one common layer in order to avoid difficulties concerning material assignments,- otherwise imported layers might be changed into independent objects during import.

Are there irrelevant bits of information in the data file? These could be, for example, remaining construction aids, or separate point information remaining in de-activated layers. It makes sense to erase this data before import and to clean the file in order to accelerate the import of data.

Is there a chance that single parts of the DTM have been referenced as blocks? In this case existing block references should be dissolved, and the assignment to layers should be tested once more.

On principle, later editing in the respective 3D visualization program is possible, but this usually involves a much greater effort. Furthermore, the data import of a DTM requires a lot of patience because it may take a long time, depending on the size of the model.

A further aspect is the matter of coordinates. In this respect a shifting vector will simplify the editing.

Geo-information necessitates huge numerical values, depending on the location of the shot, because the coordinates of the separate objects and point information have to be given as Gauss-Krueger- coordinates.

These "huge" coordinates are not a problem for a terrain modeler programmed with Double-Precision (64 bit), or for a CAD program. However, for a 3D environment such as, for instance 3ds max, which has been programmed in Single-Precision (32 bit) until release 8 (release 9 is now programmed with 64 bit), this may be a huge handicap, depending on the circumstances.

The "huge" numerical values of a terrain model may lead to strange effects when using cameras, or when doing transformations of any kind, or animations.

For this reason it is important to keep the "data values" small. Although this can be done in the respective 3D program, one has to keep in mind that deformations of the geometry can already occur here. That is why it is recommended in any case to take care of the reduction outside of the visualization software.

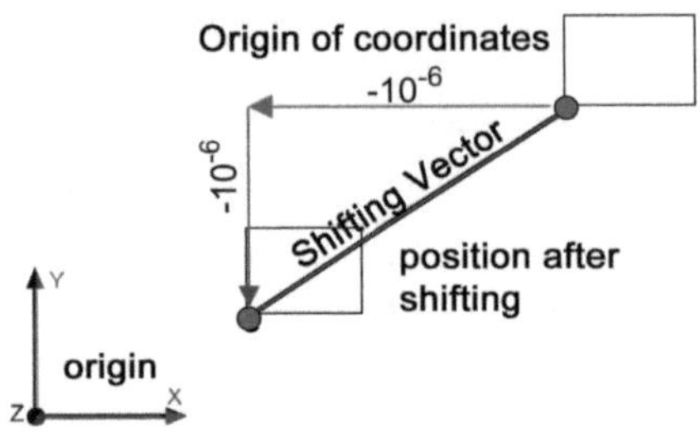


Fig. 20. Shifting of objects in the direction of the origin reduces the required memory.

The reduction of values is attained by shifting the terrain model in the direction of the origin of the coordinates. It is important to note this shifting vector in order to facilitate a later import of further information given by the client, such as information with respect to buildings etc.

### hint Optimal Data Import

*There is actually only one optimal import format for visualization programs. This is the "in-house" native data format. However, this can hardly ever be achieved. For this reason one should try to use an interface which is able to transport geometrical information with a minimum of loss. When in doubt it is worthwhile contacting the manufacturer or consulting various discussion forums.*

The following visualization examples are at times very AutoDesk-specific, because some of the given functions only work with the interface of AutoCAD and 3ds max.

These examples have been specially marked. All other examples work similarly to 3ds max but with different visualizing tools.

### *DTM import of a DWG file*

The following example of a DTM was created with the help of the AutoCad based application Civil3D. This software uses its own internal data format for describing and administering terrain data. However, it includes a direct AutoCad integration and thus enables the construction of CAD data in DWG and DXF format. DWG (**D**ra**W**in**G**) being the binary data format of AutoCad, DXF (**D**ata e**X**change **F**ormat) is the import/export interface in ASCII format.

**Fig. 21.** Example of a DTM in an AutoCad environment (Civil3D)

The following steps have been processed before exporting the file:

1. All objects belonging to DTM were shifted to one layer.
2. The order PURGE was executed to remove all unnecessary elements.
3. Construction elements such as traverse lines and irrelevant reference points were removed.
4. The remaining object was shifted by 5.400.000 x 3.400.000 units in the x and y direction towards the origin of the file.

The subsequent import[5] in 3ds max turns out to be correspondingly simple.

*FILE • IMPORT–REPLACE CURRENT SCENE*

The previous settings of the import dialogue window remain. It is important to keep the option *COMBINE OBJECTS BY LAYER* active.
The relevant breaking or caving lines which are important parts of the model remain intact and are completely integrated in to the constructed triangles.

[5] Another alternative is the option DWG linking. However, this works only with 3dsmax from version 6 onwards, whereby an external reference to DWG needs to be made.

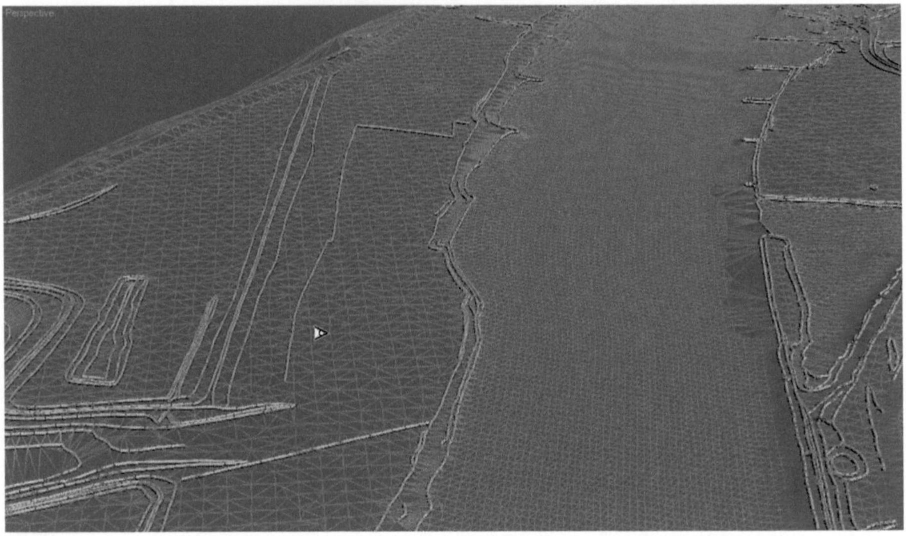

**Fig. 22.** The imported DTM. For a better illustration the fractured edges extracted from the original file were added (thick contour lines)

**Tip** **LandXML**

*Beginning with version 7, 3ds max specifically offers the possibility of importing LandXML formats. This format depicts the topology of a TIN in a succession of points and elements. This format is optimal as an interface for TIN because all information regarding breaking lines is maintained. LandXML is an Open Source Format which is supported by many GIS products and manufacturers.*[6]

[6] http://www.landxml.org

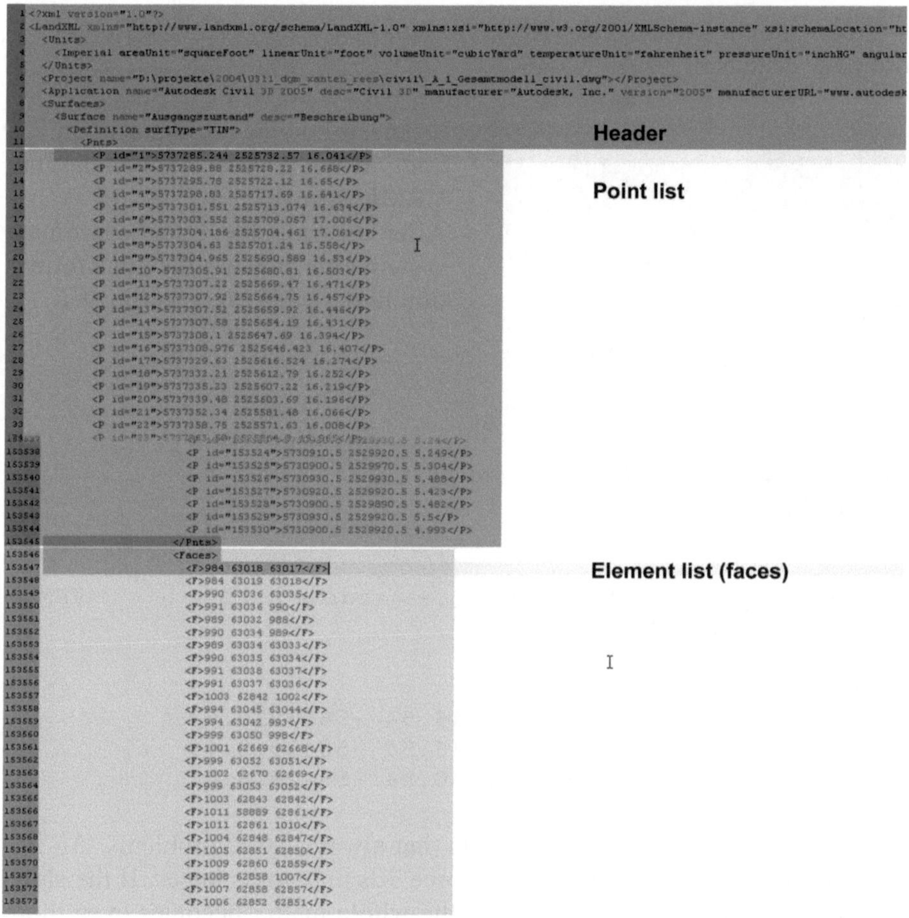


**Fig. 23.** Example of a DTM in LandXML format

## *DTM as a VRML File*

If a data file has been given in the form of a VRML file (*.WRL), this has the advantage of taking over any already existing mapping information into the VRML format. Similarly to the import of an existing file in CAD format, a VRML file can be imported into nearly any 3D program without difficulty. The decisive advantage of VRML is the fact that the data can be looked at in advance and controlled with the help of common VRML viewers[7].

[7] VRML-Viewer – an overview can be found under http://www.web3d.org

Take note, however, that VRML files may be arranged in a different coordinate system.

Fig. 24. DTM as VRML-File in Cortona VRML Viewer

The following example is an extract from a file converted into VRML format.
After entering the CAD coordinate system one notices the transformation into the VRML coordinates:

```
#CAD x,y,z = x*,y*,h*      -> turned to ->      VRML:
x',y',z' = x*,h*,-y*
.
coord Coordinate { point [
               2568885.38 46.54 -5642910.73
               2568885.34 46.54 -5642910.69
               2568886.29 46.54 -5642908.87
               .
```

But here again, it's the little things that always cause problems. An import of huge polygon numbers may force 3ds max to its knees. If the shifting vector is not taken into account, the whole matter becomes even more complicated. Importing the file requires a lot of patience. In this case you can use a ploy. If it turns out to be impossible to get to the shifted file, the problem may be solved quickly by applying an ASCII editor.

Example:

The element coordinates are distributed in a range of X/Y 2.560.000 /5.640.000:

```
coord Coordinate { point [
               2568885.38 46.54 -5642910.73
               2568885.34 46.54 -5642910.69
               2568886.29 46.54 -5642908.87
        ]}
```

To reduce the value of these numbers remove all values _256 and -564 with the option "Search/Replace". The result could look like this:

```
        coord Coordinate { point [
               8885.38 46.54 -2910.73
```

```
            8885.34 46.54 -2910.69
            8886.29 46.54 -2908.87
    ]}
```

One could of course also create a little application in Delphi or Visual Basic for the genuine transformation of coordinates. The proposed method seems easier. However, when importing a VRML file, the user has to deal with the fact that the originally coherent grid is broken up into single elements (this can also happen without applying a transformation).

You have no choice but to adapt the mesh and to connect all accessory parts to the first object of the mesh.

Similar to the direct import via CAD interface, VRML also has the advantage that the imported triangles represent the fractured edges integrated in the DTM. Thus a correct representation of the original works out perfectly.

## Import of Triple Data (XYZ)

When the data of a terrain model is only available as point information, there are different ways of importing it into the respective 3D program. For most tools, importers for ASCII data are freely available.

If you are working with 3ds max, for instance, a small free plug-in by the manufacturer Habware[8] will be helpful. The DEM23DS MAX (Terrain Mesh Import) Utility allows a Delaunay Triangulation of any triple data and generates a workable net directly in 3ds max.

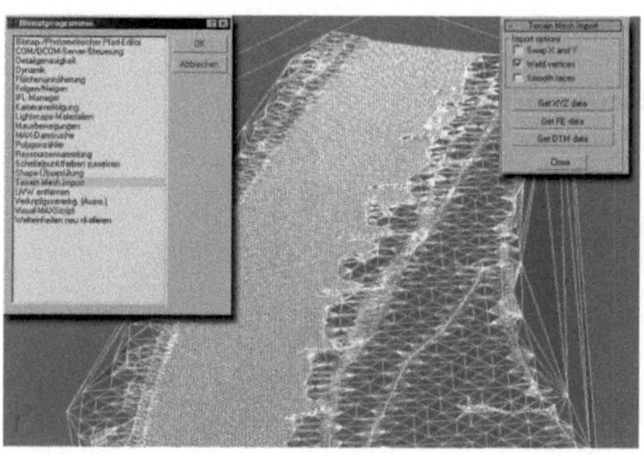

**Fig. 25.** Screenshot of the model directly triangulated in 3ds max

However, when triangulating point data anew, as a rule there is no possibility of saving information on breaking lines, so that these cannot be taken into account when meshing the point triple.

For a quick overview or for a qualitative representation of point data, the above mentioned triangulators provide an excellent service.

---

[8] http://www.habware.at

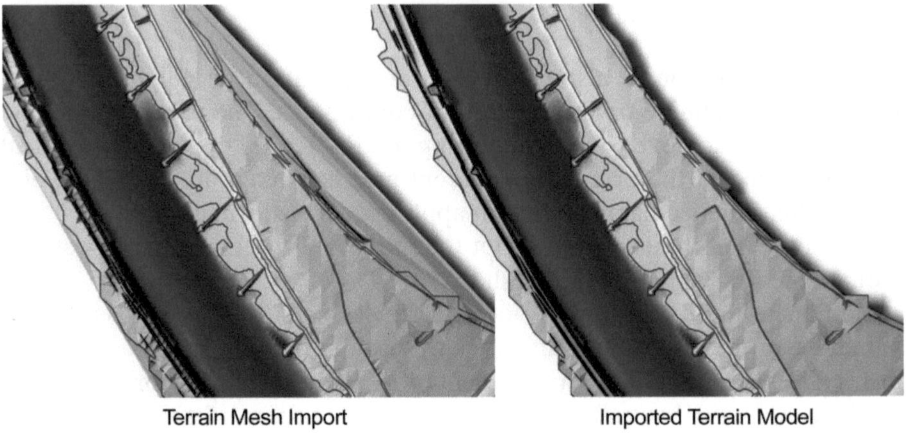


**Fig. 26.** On the left the DTM which has been triangulated in 3ds max with Terrain Mesh Import; on the right, the DTM directly imported via CAD interface. Both sets of data were first shifted towards the origin.

The first difference that catches the eye is that the CAD import has left the surrounds of the model detail intact. In the newly triangulated model the framing was ignored, which requires a new adaptation. When examining the single elements one does not notice much of a difference, but when putting both models on top of each other the differences in representation become more obvious.

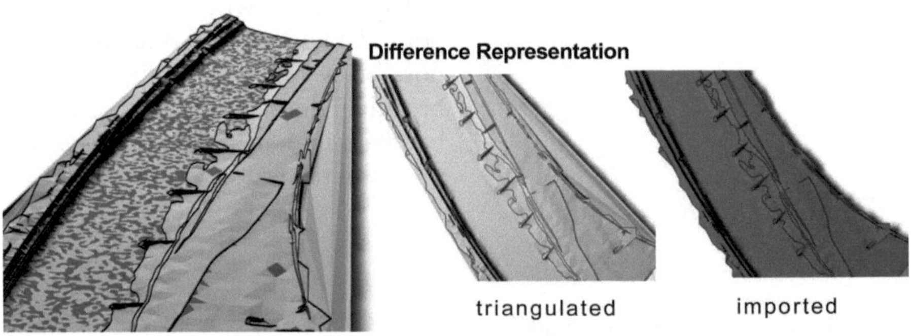


**Fig. 27.** Difference representation of both models when placed on top of each other

## Import of a DTM in DEM Format

Another possibility of importing digital terrain models into 3D-applications is the use of the raster DTM format DEM (USGS Digital Elevation Model). It is interesting to note that the LandXML importer which has been integrated into 3ds max since version 7.x can also interpret USGS formats, but this procedure is prone to mistakes.

Another very useful import tool from the house of Habware is an importer called *DEM2MAX*. The plug-in is able to read and edit USGS-DEM data.

After installation of the plug-in via copy into plug-in file of 3ds max or by later call up via loading *PLUG-IN MANAGER • LOAD NEW PLUG-IN*, the DEM format can be directly selected via *FILE • IMPORT*.

The given raster width can be adjusted afterwards and, similar to the use of the MultiRes-Modifier, facilitates a tailor made solution to the special demands of a particular representation. Furthermore, an automatic Multi-Sub-Material is created which automatically enables the coordination of height-coded information referring to the different height levels.

DEM-Import with Multi/Sub Object-Material

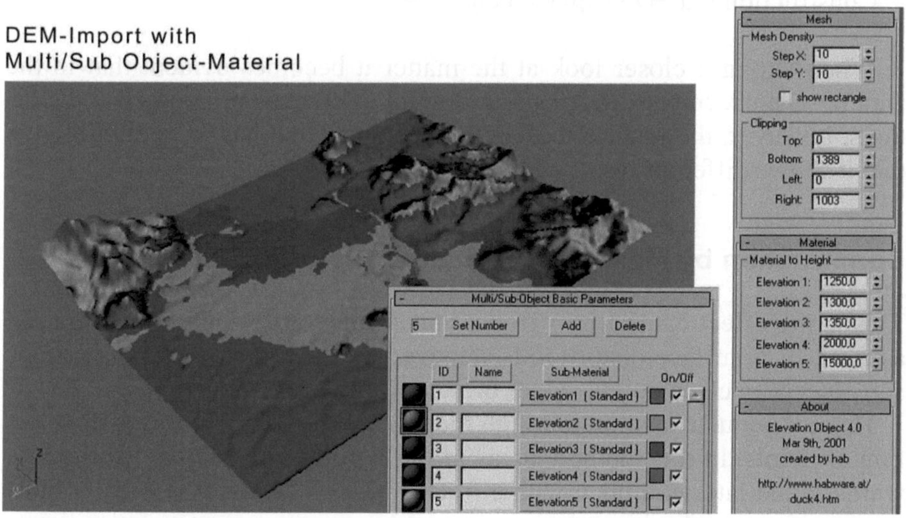


**Fig. 28.** DEM import and processing of height coding via automatically generated Multi/Sub Object-Material

However, there is a disadvantage to the DTM format. Breaking lines cannot be represented, and so the viewer has to get used to the fact that there is a distortion in this information. On the other hand, there is a lot of topographical information available for free downloading in DEM format, which gives the 3D user quick access to real geo data.

**hint** **Data Converter**

*At present, Okino's Polytrans appears to be the only pure converter which is able to create the USGS format as well.*[9]

## Building a DTM for Visualization Purposes

After having dealt with the import of existing data, let us take a quick look at some possibilities of building imaginative terrain model, using 3D visualization tools. There are three main methods of drawing up a terrain surface. These are:

1. Construction via geometric distortion
2. Use of the terrain object (only in 3ds max)
3. Construction via 3D Displacement.

When taking a closer look at the matter it becomes evident that in the end 3D Displacement procedures are nothing other than geometric distortions; however, the main effect on the geometry that has to be modified is that it shows different results.

### Construction by Means of Geometric Distortion

This is the effect caused by various transformations on any object treated as a mesh. A surface distortion is achieved by the transformation of single sub objects, such as points, edges or faces (polygons).

Here, a popular method is rough modeling with the help of a soft selection of points. In this method, areas of mesh points are (mostly) shifted towards Z and later equipped with a fractal noise. The result is usually quite an attractive terrain object.

[9] http://www.okino.com

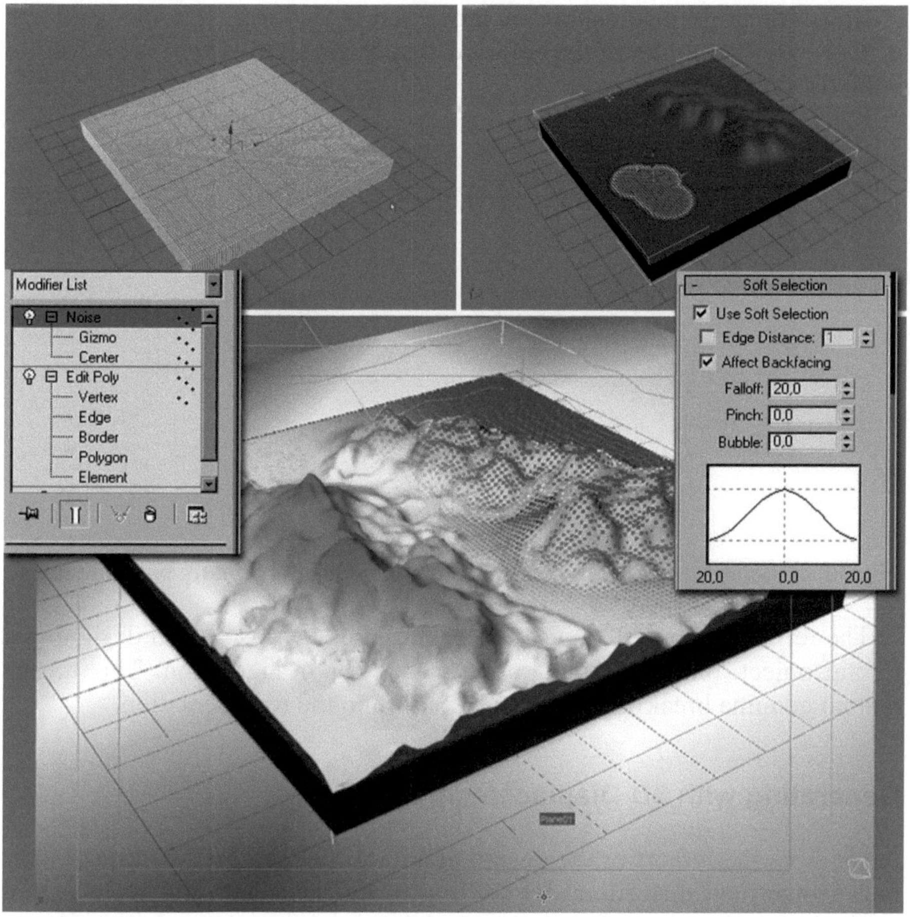


**Fig. 29.** Geometric distortion of a box by manipulating single mesh points (Edit Mesh Modifier in 3ds max) and later adding some noise ratio

## Terrain Compound Object

A very special way of drawing up topographies in 3ds max can be found under the menu *CREATE • COMPOUND OBJECTS • TERRAIN.*

Hidden behind this function is a triangulation mechanism which triangulates a terrain based on lines (shapes). The procedure is simple: at least one line has to be chosen to activate the option Terrain.

Thus it becomes possible to construct, in 3ds max, quickly and simply, "line based" terrain models. The advantage of this item lies mainly in the fact that later on further lines can be integrated into the existing mesh.

However, point data cannot be processed in this way. Another problem is, that only the points of the selected shapes are used. There is no way to build breaking lines.

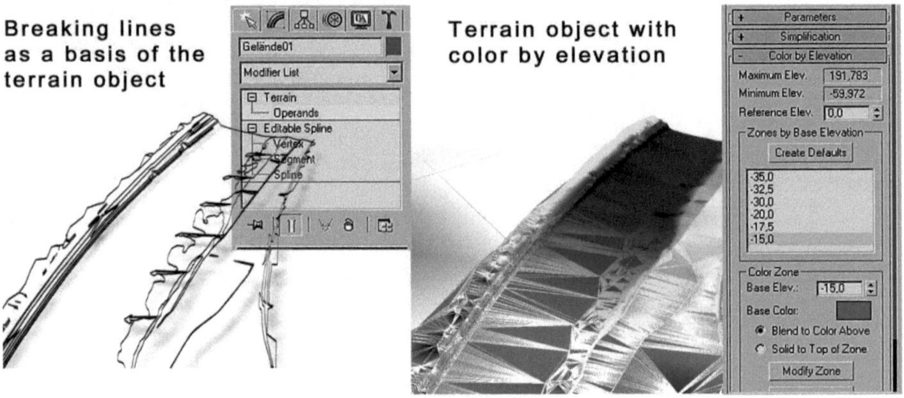


**Fig. 30.** Building a line-based Terrain Compound Object based on breaking lines

The term “line-based” is used in parentheses because when designing a terrain object, only the points can be used for triangulation. This means that no genuine breaking lines can be built into the representation.

## Generating with 3D Displacement

Apart from the import or the design of a terrain model on the basis of a direct geometrical distortion, 3D shifting or displacement is a widely used method of designing digital terrain models.

Programs like Terragen or plug-ins like Dreamscape (3ds max plug-in) make use of the possibility of assigning information on elevation on the basis of a grayscale picture. To any pixel, or rather to its particular shade of gray a specific height is assigned. The lighter, the higher and vice versa; this system of surface representation allows a quick and highly efficient design of terrain models for visualization.

However, this method of a height-coded representation is rather vague, for when using 256 shades of gray one cannot achieve highly differentiated details. But this is more than sufficient for a qualitative interpretation of an existing model or for the design of a model for the purpose of free design. The following example may illustrate the procedure:

In Photoshop (or any other program for creating pictures), a height-coded representation is drawn up. This is based on an image in shades of gray whereby the elevations have been generated with the aid of sign and painting tools of the picture publishing or paint program.

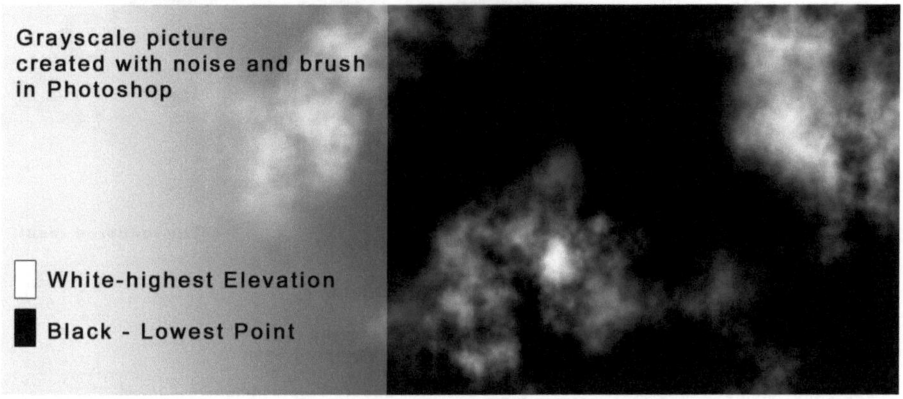


**Fig. 31.** Creating a hypothetical map in grayscales for the definition of elevation in a terrain model

A box, a plane or a Editable Polygon is drawn up in 3ds max and assigned the modifier DISPLACE. Apply the grayscale component of an image to generate a displacement. Lighter colors in the 2D image push the geometry more strongly than darker colors, resulting in a 3D Displacement of the geometry.

One can of course also use any map directly in 3ds max. If a map like noise is loaded in the material editor, this can be pulled directly via "Drag and Drop" onto the object.

In the same way as pixel graphics or procedurally generated gray shaded pictures are used for the distortion of geometrical data, picture data of all kinds can also serve as a basis for materials/textures. Textures can give you additional information concerning the characteristics or the elevation of a terrain model, and they are another very important aspect for modeling and visualization of terrain data.

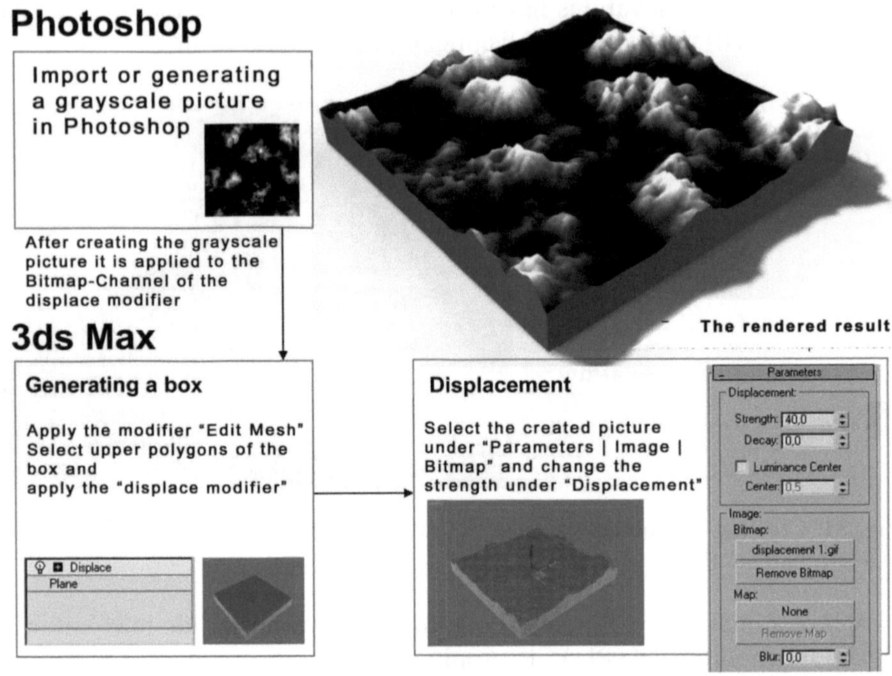


**Fig. 32.** Step-by-step procedure when using a displacement map

## Materials

The materials used define the appearance of a 3D landscape in a very decisive way. A suitable material will support and emphasize the form and shape.

But a medium can also dominate the geometry and divert the attention away from it, in a positive as well as in a negative way. Materials are describing the properties of the appearance of a surface. They are composed by maps for textures and reflections, refractions and the like. Materials are generally composed and adapted in a separate material editor.

## Maps and Mapping

It can safely be assumed that there are few areas which can make life as difficult for a beginner in the field of 3D modeling as "materials" and the related syntax.

Is this a map, or is it a texture, and what is the difference between the two? And then there are such terms as shader, procedural maps and ray-trace materials. In the beginning the whole matter is difficult to understand. It helps to clearly define three fundamental terms, and then the rest is much easier to grasp.

Material – This term defines everything that will be put on the object (mapped). A material has different parameters and is usually composed by different maps (or channels). There are maps for the diffused representation of colors (diffuse), reflection, elevation and relief (displacement, bump), and some more. It can be noted:

Map - is a sub part of material!

Shader - defines the method for calculating the materials for rendering.

Maps are accountable for the nature of the surface (e.g. a wood grain texture), the shine of the surface (glossy colors), the appearance of the surface relief, whether a material is only transparent in some places (opacity), and so on.

A map can be a picture, in other words a bitmap (pixel formats like JPEG, TGA, BMP, PNG, etc.) or a so-called procedure map, like for instance noise.

A procedure map generates patterns and structures by way of mathematical calculations, and it is independent of the resolution. This means that the quality of the map does not shrink with increasing distance of the camera from the object.

A bitmap (pixel picture), on the other hand, is completely different. A bitmap consists of a defined and limited number of pixels (picture points). You have probably come across the effect already: You zoom in on a digital picture while working with a paint program, e.g. the JPEG file in your digital camera. The further you zoom in, the bigger the pixels become. You can see the squares growing and the quality of the representation decreasing.

## Material Basics

Here are some basic definitions, before tackling the details:

**Normal Vector**: The normal defines which side of a surface or object is to be rendered, i.e. it is the exterior side. Thus the normal also indicates to which side of a surface a material will be attached.
**Specular Level**: The intensity of the reflection of a surface depends on its characteristics. The intensity of the reflection is controlled by the density of the specular level that defines the brilliance of reflection by adding shades of gray. White is a full reflection; black means no reflection at all.
**Glossiness**: This describes the ability of a material to reflect and depends directly on the density of the specular level. If this is zero, you can manipulate the glossiness as much as you will, but nothing will change.
**Tiles**: An extremely important matter when using images as materials. When tiling each picture will be put at the edge of the previous one (in all directions) until the whole surface is covered.
·**Maps**: Picture data which is ascribed to materials, e.g. to define the surface, roughness, or opacity. Maps may consist of pictures or procedural information (checker board, noise, marble).
·**Material library**: This offers the possibility to administer constructed or existing materials and maps.
·**Opacity map**: This map defines the transparency of a material and enables the manipulation thereof.
·**Bump map**: This map is responsible for the roughness of a surface. There is always a gray shading underlying a bump map. If you insert a color picture, this will be reduced to shades of gray.
·**Shader**: A shader defines the method of the basic shading parameters and their calculations when rendering.
·**Diffuse colors**: When speaking of mapping or texture, one often thinks of the diffused color or the diffused color map of a material. This is the visible surface when an object is illuminated. Very often the diffused color is connected to the surrounding color.
·**Ambient color**: This defines the representation of those parts of an object that are lying in the shade.

### Tip Filing Materials/Collection of Resources

*Create your own files to administer and label your maps and picture data. Make sure that this data is regularly saved. It forms part of the project data and a loss can hurt badly.*

*In 3ds max you will find a small tool, among the service programs, called resource collector which copies all textures of a file into a defined directory.*

Depending on the method of visualization, the choice of materials is decisive for the final picture message of a landscape visualization.

In aerial photography or when assigning mapping materials the problem of geo-referencing surfaces appears again and again.

Normally, aerial photographs are edited as TIF, rectification is applied and they are fitted true to scale. Additionally, the coordinates and the alignment of the picture are delivered in a separate text file. The corresponding programs for designing a DTM and for GIS applications are able to interpret this text file and to recognize this picture as an item to be used as a map.

There is no such possibility in purely 3D visualization tools like 3ds max. Here, the only way out is to draw up a map manually and to adjust it to the corresponding topography, which involves a big effort.

If you have to visualize many terrain models, it is recommended to take a look at products like the World Builder or World Construction Set because here the possibility of including geo-referenced picture information is a built-in part of the functions.

### Tip A useful Tool

*A very nice tool for integrating geo textures is the plug-in Wavgen from Wavgen Technologies. This program is available as a plug-in for all current 3D programs and enables the loading of big aerial pictures and Geo-Tif.*[10]

Very often natural picture material is used for producing a material. An incredible variety of pictorial material for any subject can be found on the

[10] http://www.wavgen.com

market, be it surfaces or textures or backgrounds for any picture mood. You will find anything your heart desires on the free market. This commercially sold data has normally been fitted to the requirements of a 3D visualization. This means it is tileable, the distortion has been rectified, and it is usually available in a variety of resolutions.
But if you want to take and/or fit the pictures yourself, there are certain points you will have to consider. Important aspects for handling and drawing up materials are:

Angle of the shot
Picture Resolution
Scanner
Grid effects

### Shooting angle

When taking your own pictures for use as textures or maps, watch out for the right positioning of the camera, namely orthogonally to the object to be photographed.

In background pictures you can bend the rules a bit, but when a photograph has to be fitted to an object, oblique takes can lead to very strange results.

### Image Resolution

When using images as textures or maps it is important to use a suitable resolution. Depending on the specific use, e.g. for an object in the background of your representation, a picture in PAL-resolution (768 x 576) is quite sufficient. But when you plan to do a close-up you quickly come to a dead stop. The picture data, which was great for the overall scene, looks pixely and square all of a sudden when used for an object close by. There is no general rule for the "right" resolution.

The attempt to put in all pictures with the highest possible resolution will certainly provide the desired quality, but one should keep in mind that one is dealing with limited resources. Any 3D visualization will quickly come to a halt when the complete memory of the computer and the graphic card is wasted on dealing with unnecessarily large texture files.

In this case only a lot of experience acquired by trial and error and supported by a few rules of thumb can help. For sure, it is better to use a picture with a resolution that is too high rather than too low, because it is always possible to reduce the number of pixels afterwards. Unfortunately the opposite procedure remains impossible.

### *Scanner*

Although the scanner is replaced more and more often by the use of digital cameras, it still is and remains an important tool for the work station of a 3D user. Often pictures from magazines, paper photographs or information flyers become material for a landscape visualization via the scanner.

The sets which can be purchased in many supermarkets are quite sufficient for a quick scanning. The set should at least be equipped with 600 dpi as physical resolution[11]. 1200 dpi is nowadays standard.

### *Grid Effects*

When using a scanner there can be undesirable grid effects. These effects are the result of the resolutions used in printing (usually 300 dpi), which are accepted without any problem by the human eye. The scanner, however, ignores these effects which often results in an image with Moiré-like effect. These grids can be removed without any problem by a picture program like Photoshop, which uses a Gaussian soft focus.

## Materials with Color Gradient for Height Representation

When visualizing terrain models it is often important to gather, as quickly as possible, the relevant information and the distinctive features of the topography. It can be very useful to implement a height-coded representation. A color scale, adjusted according to the requirements, can give a quick impression of the important elements of a terrain (here again, the limitations of a black and white print are obvious). There are two ways of achieving a height-coded representation of the model, and in most 3D programs they work in similar ways:

Use of a picture (bitmap/pixel graphic) or
Designing a procedural color gradient.

Admittedly, both kinds of representation are not really suitable for use in a presentation for the public, but they can help accelerate the decision process amongst experts.

---

[11] Physical resolution means that the resolution has not been achieved by a later extrapolation of pixels, but that it contains the above-mentioned resolution by the corresponding optical features.

### *Bitmap*

If, within a visualization, a predetermined color scale is used, the designer can get quick results by applying one of the respective bitmaps. The color scale can be saved as Bitmap (TGA, JPEG, TIF), the corresponding polygons which will be part of the color scale can be selected, and the correspondingly assigned UWV mapping can be turned by 90° and adjusted to the selection.

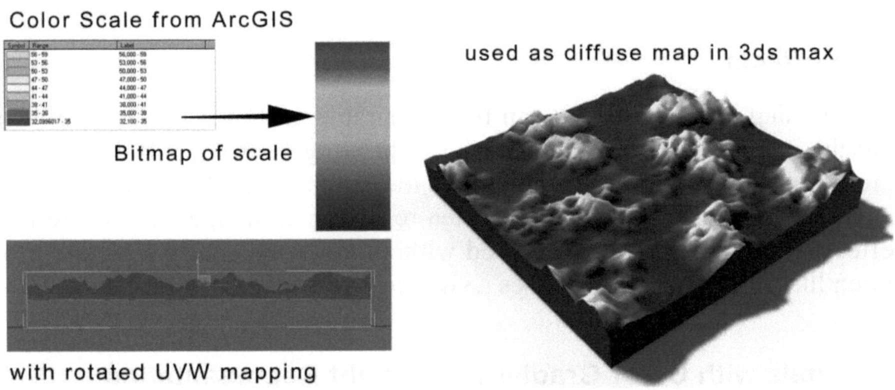


**Fig. 33.** Using a color scale generated by Argos as a bitmap in the diffuse color channel in 3ds max

The advantage of using a bitmap is that this allows the quick and uncomplicated use of any other height scales designed for previous projects, via the mapping function of the respective 3D program. Thus the fine adjustment becomes unnecessary in a new design. And anybody who ever had to transfer, in ArcGIS, the height coded color setting into another model will admit that, truly, most problems arise when dealing with details.

The disadvantage when using pixel graphics becomes evident in the question of resolution or, in other words, how near the camera has to zoom into the represented terrain. As always in pixel graphics, the infamous staircase effect (aliasing) will surface at some point.

Thus, if a new height coded representation needs to be generated, it is recommended that you use a procedurally generated gradient ramp.

### *Procedural Color Gradients*

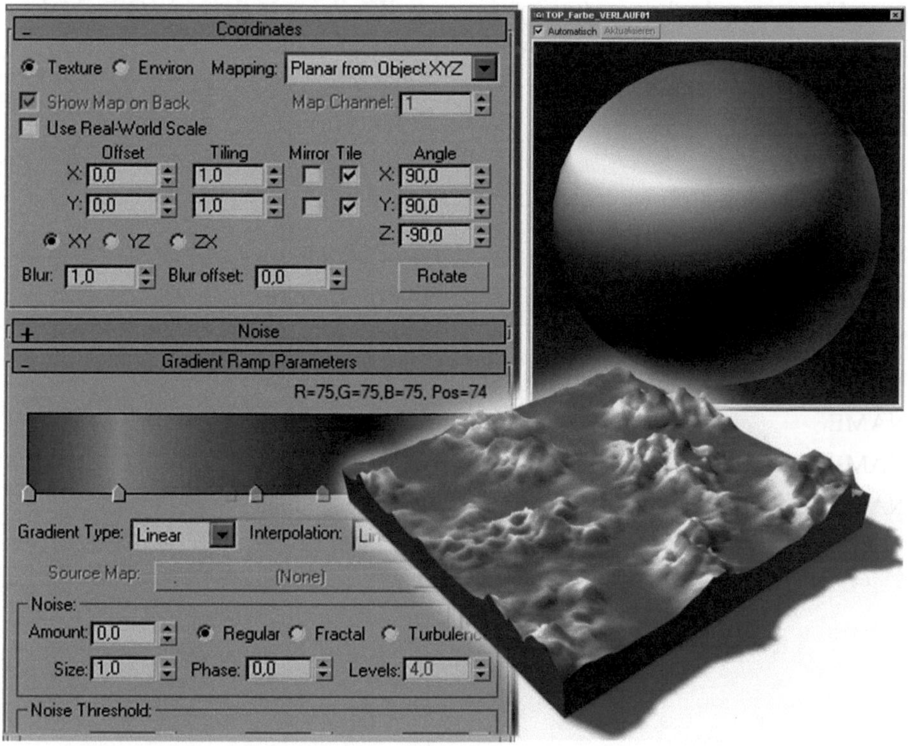


**Fig. 34.** The map gradient ramp enables you to draw up color coded height information procedurally and quickly in 3ds max.

In a procedurally generated height coded representation the color gradient is set up mathematically. The advantage of this is that the quality of the representation does not decrease with increasing distance of the camera to the object. If height coded representations need to be newly generated it is strongly recommended to furnish these with a procedural material. As a rule, such a procedural material is built as a map into the diffuse channel of the material.

### Tip Labeling of Maps

*It is recommended to keep a watchful eye on the naming of picture data, in order not to lose control over all maps and bitmaps. However, for bitmaps you should try to choose file names without special characters like ", " (/, +, #). Depending on the use in rendering or exporting, names with special characters can pose difficulties, especially in systems like Linux, which are found in many platforms. Apart from that, Windows is the only OS which accepts the use of small or CAPITAL letters.*

*Here is a suggestion:*

**Table 1.** Naming different materials

| Name of Bitmap | Kind of Map | Name addition |
|---|---|---|
| NAME | Diffuse | NAME_C(color) |
| NAME | Specular | NAME_S(pecular) |
| NAME | Opacity | NAME_O(pacity) |
| NAME | Bump | NAME_B(ump) |
| NAME | Reflexion | NAME_X |
| NAME | With Alpha-Channel | NAME_C_alpha |

## Mixed and Composite Materials [12]

**Fig. 35.** Landscape with top/bottom material. Depending on different parameters such as the normal alignment, the landscape is covered in different ways by composite materials.

### *Fields of Application of Blend Materials*

The principle is simple. An existing material is cross-faded by a second (or third, etc.) material according to varying criteria. Depending on the applied blend material, these criteria can be aspects like

- Height/elevation
- Sloping angle
- Contours
- Spots (not an even cross-fade but a spotty one)
- Simply a picture with transparency information (e.g. alpha channel or a separate opacity map)

---

[12] Often the term "composite materials" is used for mixed and composite materials. Fade-over materials are called blend materials

### hint Opacity and Transparency

*Opacity enables an overall control of the imperviousness of a material to light. Low values correspond to a low impermeability, high values to a high resistance to light. With this parameter you cannot simulate refractions, i.e. the light refraction by different materials. In order to simulate light refraction one usually applies a refractions map.*

*Transparency is the opposite of opacity. It defines the permeability of a material to light as opposed to its opacity, which defines the impermeability of a material.*

*100% opacity = 0% transparency*

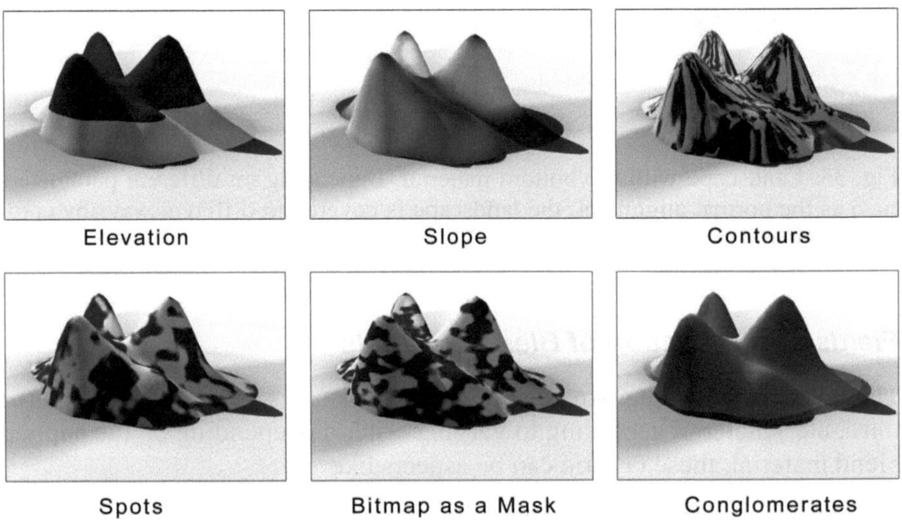


**Fig. 36.** Use of a Mental Ray material for the representation of terrain surfaces

The surfaces shown in Fig. 36 were drawn up with the Lume-Shader Landscape of the Mental Ray Renderer. If this renderer is not available for application, there is, of course an easier way. Usually, a standard blend material always offers the possibility of fading two materials into each o-ther. The following example in Fig. 37 was drawn up with a bottom/top material and a simple standard material.

### *Snow-covered Mountain Peaks*

In this material, the top material covered all surfaces whose normals point upwards. The surfaces with downwards-pointing face normals were covered by the bottom material. A cross-fade effect can be applied to the transition point. This can either be a specific area or even a grayscale bitmap. If such a bitmap is used the dark areas are transparent and the light areas are opaque. The ones with intermediate values are interpolated accordingly.

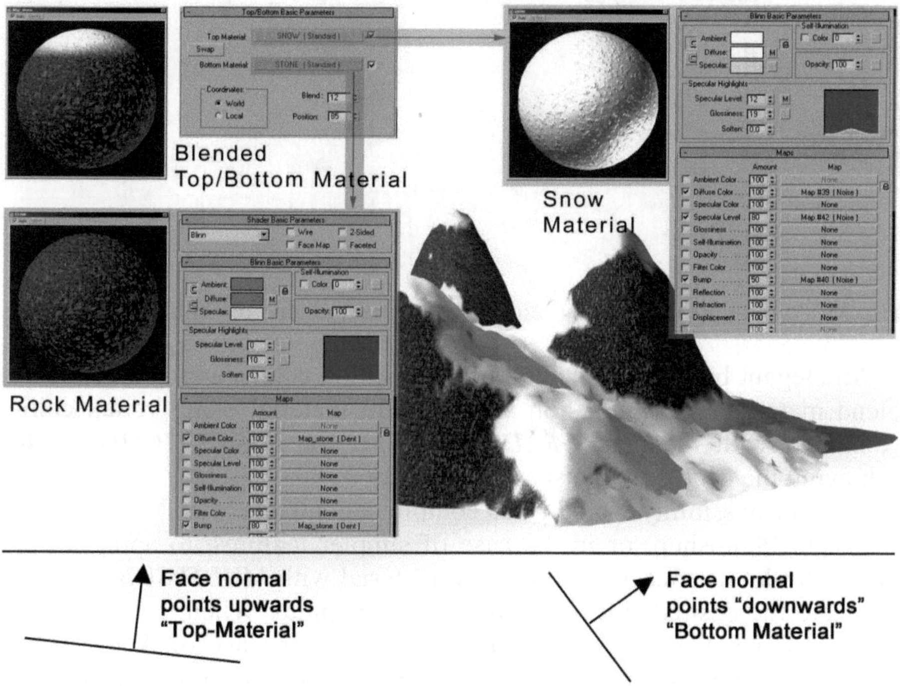


**Fig. 37.** Top/Bottom Material as a simple solution for the application of a fade-over material on the basis of the two materials for rocks and snow cover.

The alignment of the face normal does not necessarily have to point upwards or downwards. Depending on the alignment of the local object, this can point forward or behind, or to any defined angle.

### Transition Areas

A further important use of composite materials is the transition between two materials, or, to be more specific, the avoidance of sharp edges between for instance the side of the road and the adjacent grass cover.

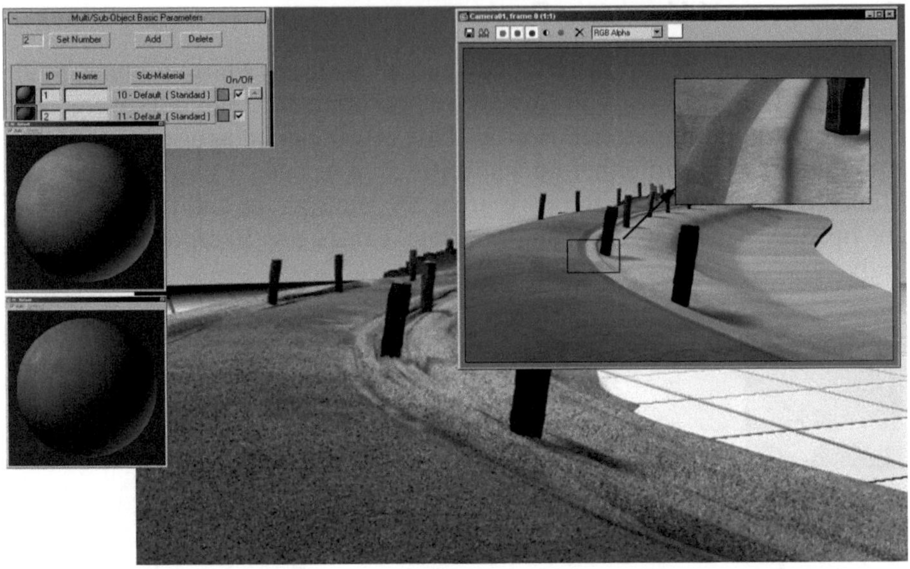


**Fig. 38.** Avoiding sharp edges (color jump) by using blend materials.

An elegant but simple method of achieving a color jump is the use of a blend material in combination with a multi material using vertex colors combined with the map *GRADIENT RAMP* or *VERTEX COLORS* for achieving the necessary transparency effects.

Another possibility would be to compile a complete "final" map, for instance in Photoshop, or in the case of simpler textures, to apply a *MIX MAP*. But let us concentrate on blend material with *MULTI/SUB-OBJECT* to create the desired soft transition, the color jump. When using data imported from CAD programs (especially at the interface of AutoCAD to Autodesk Viz or 3ds max), these are, depending on the editing of data, already imported into 3ds max with the corresponding multi materials, or respectively have been referenced via DWG linking.

**Fig. 39.** The result shows a road with an adjacent grass surface. The transition between road and terrain must NOT be sharp but soft and faded.

Of course, nearly every type of geometry can be optimized and fitted afterwards and the corresponding UVW-coordinates etc. can be added later on. But this kind of creative approach is not really reconcilable with the guidelines of a normal workflow in the office where time and money are scarce commodities. And when the data has already been fitted with multimaterials those should be used as efficiently as possible.

## *Model and Material Index (ID)*

Since real project data (import data) is usually considerably larger than needed, for instance, during a practice session, a loft extrusion object (*LOFT COMPOUND OBJECT*) drawn up in 3ds max will do.

Loft extrusion objects are two-dimensional shapes extruded along a third axis. In this case, the path defines the extrusion direction. The shape defines the cross section. It is interesting to note that in loft compound objects the material ID of the shape used for the cross section (usually given as spline) can be transferred to the surface. For the example of a road on a dyke, to begin with, two different IDs are provided.

First in the front view a shape (cross section) of the dyke is drawn up with the help of a spline. The segments of the shape which will later correspond to the road are fitted with the material ID 2, and the segments from which the terrain will be generated are fitted with the ID 1.

**Procedure for the allocation of IDs:** Select the spline, activate *MODIFY* • *SUBOBJECT SEGMENT* and allocate the designated ID under *SURFACE PROPERTIES*.

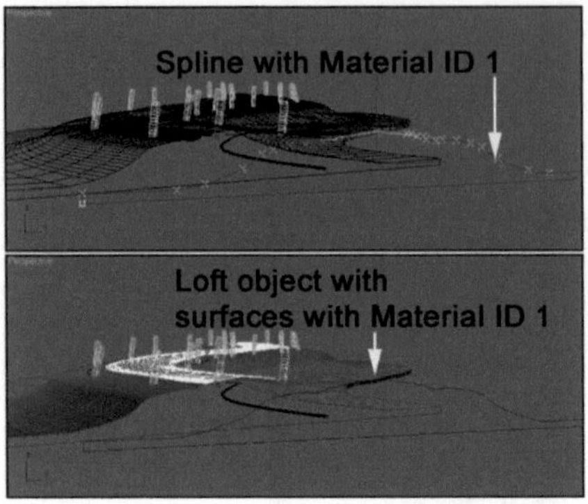


**Fig. 40.** A spline as cross-section (shape) was chosen as the basis of an extrusion object. The material ID 1 was assigned to the line segments of the terrain, the material ID 2 was assigned to the area of the road.

Afterwards a *COMPOUND OBJECT "LOFT"* is generated with the help of any path. To activate the Loft-object either the path or the cross-section has to be selected beforehand.

**Procedure:** *CREATE • COMPOUND OBJECTS • LOFT* then either select path (if shape has been selected), or select shape (if path was selected before).

Activation under *SURFACE PARAMETERS • MATERIALS • GENERATE MATERIAL IDS* and *USE SHAPE IDS* ensures that the material ID of the spline will be transferred to the respective areas.

In the present case, with only two materials, the easiest way is to use a blend material to adjust the transition between these two materials with the help of a corresponding interface. However, blend materials are not adjusted to the material IDs of an object but take effect from top to bottom. Therefore one has to watch out for a few minor but important details, such as the following.

### *Blend Material with Gradient Ramp*

One at a time, a standard material for grass and asphalt is created separately. Afterwards the two materials i.e. *MATERIAL GRASS* and *MATERIAL ROAD*, are fitted to a new *BLEND MATERIAL* and to the respective interface.

**Blend Material** – On top *MATERIAL ROAD*, bottom *MATERIAL GRASS*

**Mask** - The map *GRADIENT RAMP* as a transparency mask will keep the underlying material, colored in black, visible.

All white areas of the mask are assigned to the visibility of the top material road. A fractal noise added in the map *GRADIENT RAMP* will provide the required "material chaos", the color jump, in the transitional areas.

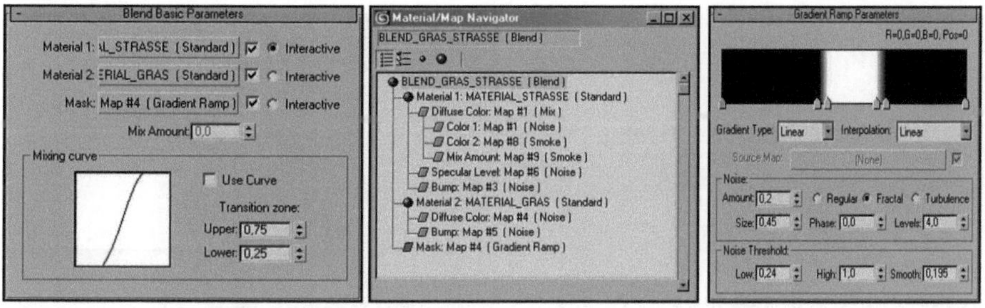


**Fig. 41.** Creating a blend material including mask by using the map *GRADIENT RAMP*

The overdrawn effect of the map gradient ramp and its effects become apparent in the following illustration.

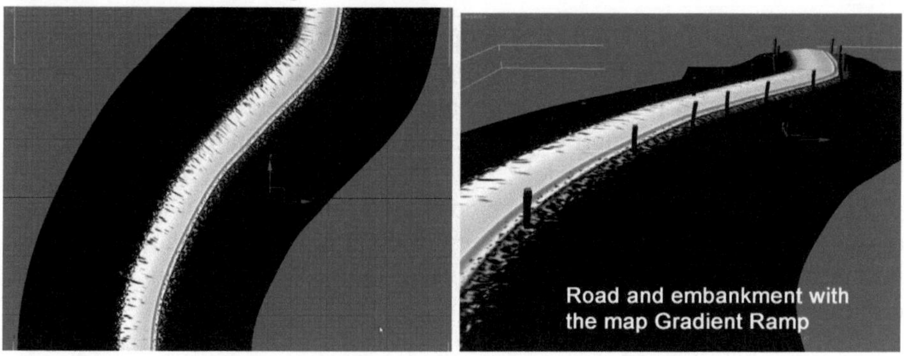


**Fig. 42.** Gradient ramp for creating the transparency information

## *Blend Material with Vertex Colors*

A loft object where the texture coordinates can be fitted without problems, as in the example above, is not always available. In this case, the use of Gradient Ramp as transparency information can, at times, become difficult. Also, instead of the complete stretch of road, sometimes only a small specific part is needed for the color jump. In such a case the use of vertex colors is a good alternative.

**Fig. 43.** The left picture shows all areas in white after assigning the modifier VERTEX PAINT. In the right picture, the effect is shown and the surface is covered with the material grass.

Instead of the map Gradient Ramp used before, the transparent area is now "painted" with vertex colors. Vertex colors can be used in a variety of ways, especially in interactive applications, to connect additional information to an existing geometrical form. These can be illumination values or radiosity solutions for example.

The 3ds max online help can be consulted on this subject but the tutorials are not really encouraging. For beginners it is worthwhile to take a look into current 3D forums[13] or to try the term in one of the current search machines.

Select the object (in the example the Loft object had already been supplied with the modifier *EDIT POLY* to change the edges of the ID) and add the modifier *VERTEXPAINT* .

In the opening flyout window, the option *VERTEX COLOR DISPLAY-SHADED* is activated and the whole model turns to white. White is the preprogrammed color of all vertices in 3ds max. In Fig. 43 the rendered result only shows the material MATERIAL GRASS, there is no trace of the road. The further procedure is simple. Select the desired color (black for transparency); define the brush size (BRUSH – 50 in the example), and "PAINT OVER" the area to be transparent later on (Fig. 45).

[13] Web boards like http://support.discreet.com/

Fig. 44. Fly out window VertexPaint

Optionally, sub objects like vertices or faces can be selected as masks. Like in a normal picture-editing program the painting is restricted to the objects selected.

The edges of such a brush action can later be comfortably fitted and softened or blended by choosing a selective soft focus *BLUR BRUSH*.

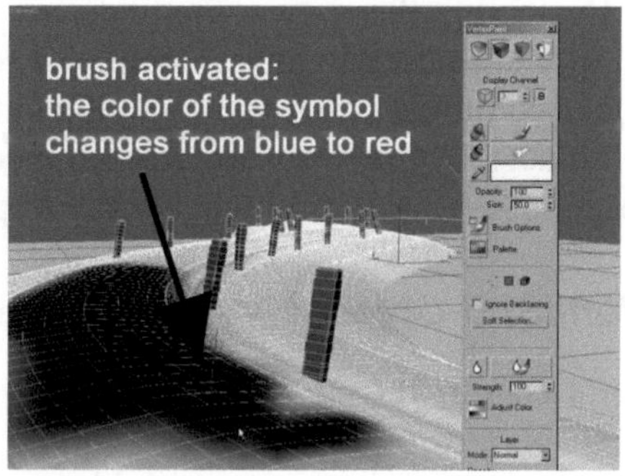


**Fig. 45.** Painting with the brush

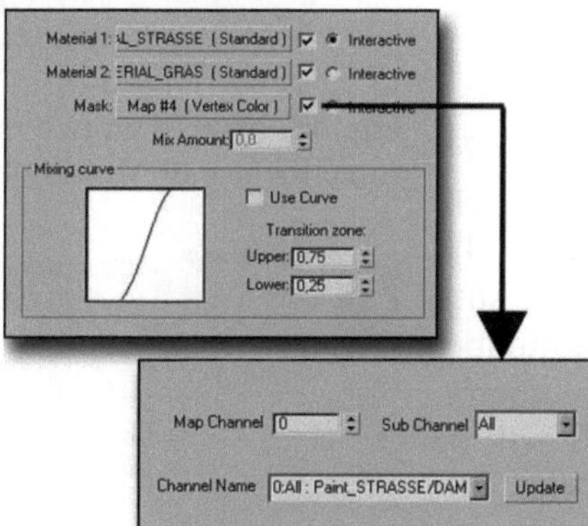


**Fig. 46.** Assigning Vertex Color as mask

But before the result can become visible when rendering, *VERTEXCOLOR* as a map has to be assigned to the blend material. (Fig. 46).

An input requirement is the setting 0 of the *MAP-CHANNEL*. If there are uncertainties concerning the use of the *MAP-CHANNEL*, the tool under *TOOLS • CHANNEL INFO* will be helpful.

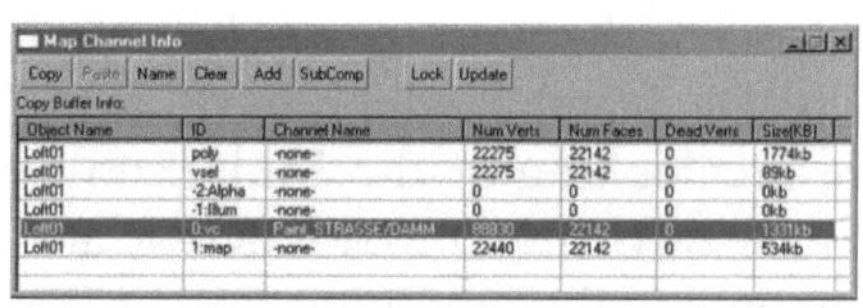


**Fig. 47.** Examining with Channel Info

Basically, these two methods are quite sufficient to create the color jump, i.e. to blend embankment and road. But when you have to edit more than two material IDs into one object the blend material is not sufficient. A reminder: If you want to avoid unnecessary work, multi materials are an essential input requirement for the import of CAD-data.

### *Multi/Sub-Object Material*

Let's assume the surroundings of the dyke or any other geometric forms that are part of the entire model have already been fitted with Multi/Sub-Object Material, and now require further sub-materials. One has to take these requirements into account. To illustrate this the surrounding area of the dyke in the above example was fitted with the material ID 3 (Fig. 48) *MULTI/ SUB-OBJECT* provides 10 different sub-materials by default (Fig. 49. The numbering of the various sub-materials corresponds to the number of the material ID of a geometry. This way, the sub-material 1 will be automatically allocated to the surface with the material ID 1, the sub-material 2 will be allocated to the surface with the ID 2, and so on.

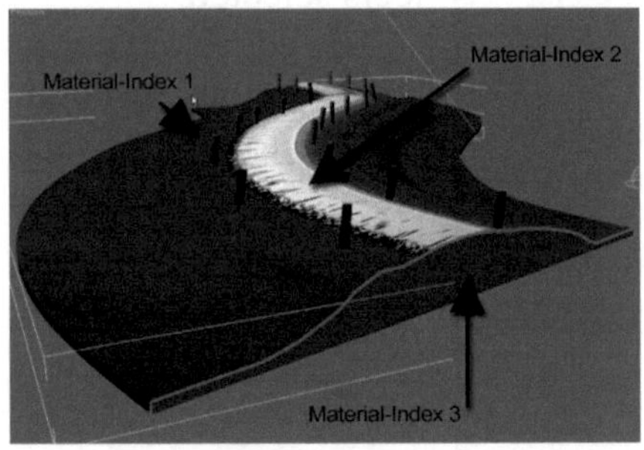


**Fig. 48.** More than two materials necessitate the use of Multi/Sub-Object

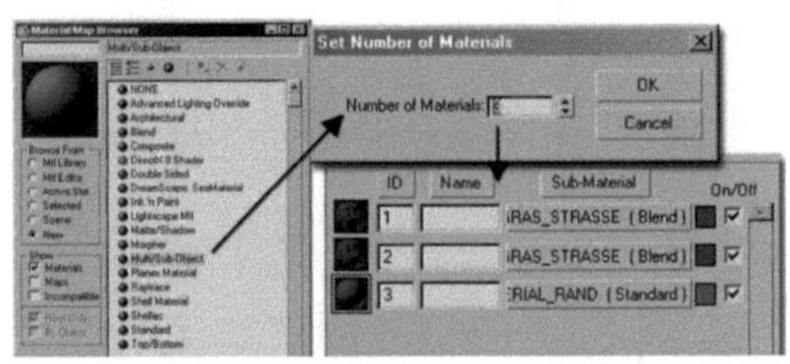


**Fig. 49.** Creating a Multi/Sub-Object Material

Thus, the Multi/Sub-Object Material is fitted with 3 material IDs, the third ID being allocated to the border areas with a blue material (MATERIAL_BORDER). The multi-material is allocated to the whole geometry.

In fact, the work is nearly complete, the problem areas having been the two IDs 1 and 2 which have been "covered" by the blend material. Now we are looking at the Multi/Sub-Object Material to let the first two IDs stay with BLEND MATERIAL and allocate the corresponding material to any potentially emerging further IDs.

The two channels for ID 1 and 2 need to be fitted with the blend material that has been created before (example see Fig. 49, and that's all there is to it. This step enforces the use of the Blend Material for ID 1 and ID 2.

All further IDs get their material automatically. In this way the Multi/Sub-Object Materials as well as the color jump are brought together in a very simple manner.

## Mapping Coordinates

If you want to assign a 2D map material (or a material containing 2D maps) to an object, this object needs to have so-called mapping coordinates. These coordinates specify how the map has to be projected onto the material, and whether it will be projected, tiled or mirrored as a "Decal". Mapping coordinates are called UV or UVW coordinates.

These letters refer to the Object-Space Coordinates[14], as opposed to the XYZ Coordinates that describe the whole scenery.

Maps are spatially aligned. If a material containing maps needs to be used on an object, this object needs to have mapping coordinates. These are indicated with the help of local UVW axes for the object.

When using procedural maps, normally no mapping coordinates are necessary, because directly at input, procedural maps are defined parametrically in size and behavior.

## Tiles

As it happens, a bitmap intended for use may be too big or too small to produce the desired surface effect. Floor tiles or wallpaper patterns are a typical example for the use of tiles. A small section of the pattern called a tile is used, which will then be repeated n-times depending on the manner of tiling. Usually the number of tiles in the U and V direction of the map is specified.

When using tiles the seamless transition between the single tiles is important. In this domain there are already many ready made textures that include the option "tileable". It is worthwhile to start an internet search and look for tileable textures.

### Tip Texture/Size of Images

*With the probable further use of interactive applications of textures in mind (see chapter Interaction with 3D data), it is recommended to keep an*

[14] Object related coordinate system, comparable to user coordinate system, as opposed to the XYZ Coordinates that describe the whole scenery

*eye on the size of the pictures. Recommended picture sizes are 2 power to n like for instance* $2^2$, $2^4$, $2^n$ *pixels.*

There is a variety of software products on the market to produce your own tileable file from a bitmap.

**Tip Alternative Picture Editing**

*The picture editing program GIMP*[15] *is Open Source and will run on nearly every operating system. It is similar to Photoshop.*

If you have to design a tileable texture yourself, the following example will show you one method:

During a walk in the countryside one comes across a dry river bed. Thanks to the digital camera a top view of the sub grade was taken immediately. The picture is to serve as a tileable texture.

The editing in this example is done with Photoshop; it can be done in a similar way with nearly any picture-editing program.

---

[15] GIMP - GNU Image Manipulation Program http://www.gimp.org

Fig. 50. The photographed riverbed

**Calculating the size of the picture**
The picture in the example has the dimensions 1300 x 828 pixels.

Fig. 51. Shifting effect in Photoshop

**Filter**
Under the menu *FILTER> OTHER FILTERS> OFFSET* you put in half the width and height of the picture. In the example these are 650 x 414 pixel. The figure shows the picture which has been fragmented into four segments.

Fig. 52. Mask for the borders

**Masking**
First the part of the picture, which needs to be edited, is provided with a mask. The mask protects the other picture areas and reduces further manipulation to the masked area.

Fig. 53. The result after editing with the stamp-tool

There are, as always, various possibilities for retouching the troublesome undesired borders. One method is the use of the stamp tool. Here a specific area of the picture is selected (by simultaneously pressing the ALT key).

Fig. 54. The result after color correction

Depending on the sub grade and the lighting while taking the picture there can be differing color areas within the picture. In this case it is recommended to apply a color correction to the extremely varying areas. Here a simple method is the use of the tool DOGE or BURN.

A quick method of controlling the result is already available in Photoshop. The edited picture is taken as a prototype and used to fill an area (pattern).

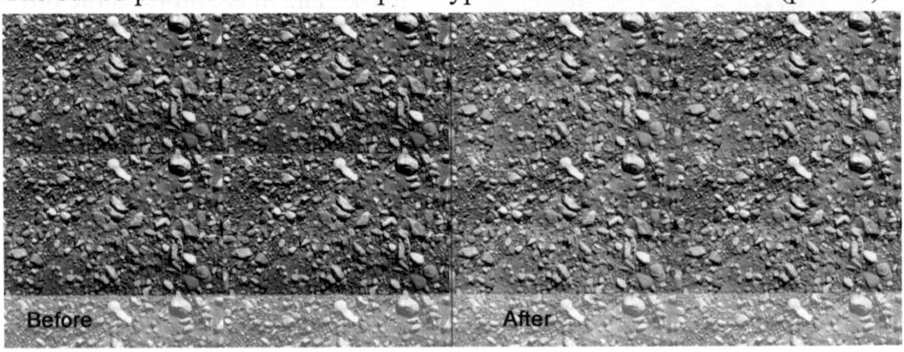


Fig. 55. In order to examine the result the original picture (left - before) and the edited version (right - after) are displayed next to each other.

In order to fix the prototype in Photoshop you proceed as follows: mark the entire picture area with *SELECT > SELECT ALL* and define the pattern via *EDIT > DEFINE PATTERN (CTRL + A).*

**Fig. 56.** A "real-life" example for a unfortunate editing of tileable textures

## Terrain Distortion

The surface of the earth, which is the subject we are ultimately dealing with, is a very dynamic thing, although the dynamic changes take place during periods of time that are beyond our perception. Faults, deformations, and volcanic eruptions with lava deposits, plate tectonics - the surface of the earth is moving and it is moving continuously. Sometimes, at the moment of a natural catastrophe, we may become witnesses of a geological event in "real-time", which we would never see normally due to our short lifespan. It is in this field that the strength of digital visualization is to be found. With the help of suitable data, geological events can be represented in fast motion, so that we can perceive and register these processes.

Apart from tectonic and volcanic activities, water is one of the strongest shaping forces in nature. Erosion or frost damage, the change of seasons, precipitation or wind, all these are components that can destroy even the toughest rocks.

Let us take a closer look at the force of erosion. When looking up the term “erosion” in Wikipedia.org, the following definition is given:

“Erosion is the displacement of solids (soil, mud, rock and other particles) by the agents of wind, water or ice, by downward or down-slope movement in response to gravity or by living organisms (in the case of bio erosion). Erosion is distinguished from weathering, which is the decomposition of rock and particles through processes where no movement is involved, although the two processes may be concurrent.”

But we should not get our hopes up: no 3D visualization program can, by itself, execute calculations on erosive algorithms, but it is possible to import and represent existing data. Apart from this, a simplified method of representation is feasible by means of internal tools. Although the scientific aspect falls by the wayside, the results are quite sufficient for a qualitative representation.

The displacement method is the quickest way of reaching the desired effect.

The basis of the model is a grayscale picture file. The desired erosion effect is achieved with the help of a picture editing software. Again Photoshop is used.

Depending on the bedrock the erosive agents either remove the soil, or smooth it out, or, when more drastic effects are caused by, for instance, lakes and rivers, the changes in the subsoil can be rather “sharp-edged”.

The option *MEZZOTINT* in Photoshop is just one of many possibilities. It certainly makes sense to study the filter functions in Photoshop in more detail, and to test some of the other functions and filters as well.

### hint Modelling Erosion in 3ds max

*If you want to study the subject of erosion in more detail, you will find a suitable product at Sitni Satis Dreamscape[16].*

[16] www.afterworks.com

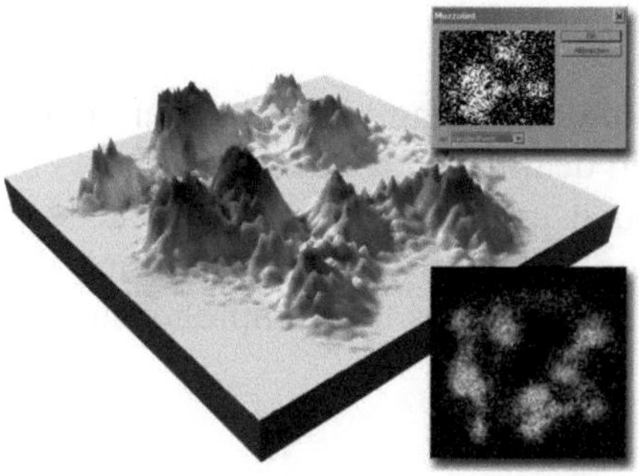

Fig. 57. Original Terrain

**Original Terrain**
Based on a grayscale picture, which was generated in Photoshop, the o-riginal terrain is constructed via displacement.

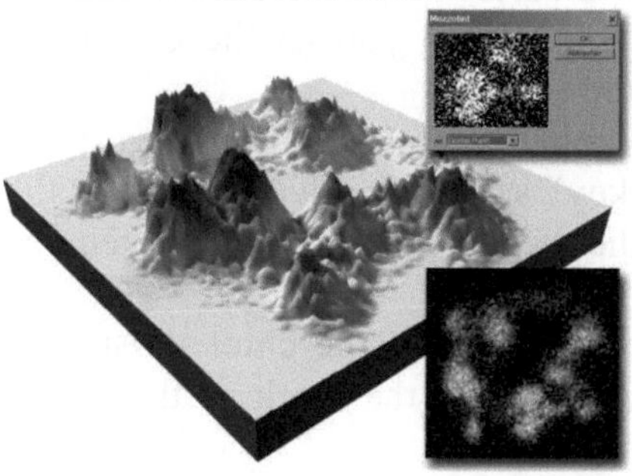

Fig. 58. Erosive Effect by means of Mezzotint

**Filter in Photoshop**
Under *FILTER > PIXELATE > MEZZOTINT...* a coarse raster can be given to the sharp-edged degradation of the terrain. The altered picture file now substitutes the original displacement map.

Fig. 59. Reiteration of the Effect

**Amplifying the Effect**
Further stages of erosion can be simulated without problems by re-peating the effect of the mezzotint filter.

## Animations

A movie sequence gives better insight into changing variables than a still ever could. That is why, especially when dealing with changing aspects of a landscape through the above-mentioned erosion, the next logical step will be the creation of an animation. The possibility of demonstrating the time component of change in a landscape facilitates the comprehension of these processes tremendously.

### A short Excursion

Have you ever heard of the flipbook? In a series of pictures a drawing is continuously changed by very small details. When you let these pictures flip, the impression of movement is created.

The term "keyframes" was originally coined in the traditional world of animated movies. Keyframes were the "key" scenes that were done by the chief illustrator. The scenes in between these crucial scenes were delegated to junior artists.

In 3D programs you only have to do the work of the chief illustrator. The programs will fill in the gaps; they take over the work of the junior artists.

Anything you animate in any way is done by keyframes. In the present case this means that for every change that is to be done to an object over time, the respective key at a specific point in time has to be created. This key then represents settings like for example position, transformation, visibility, etc.

Thus animation via key framing always has a variable over time. It is the same with a movie: a series of pictures runs by at a defined speed - the human eye notices movement, because the speed is usually too fast to recognize single pictures.

For a possible animation of terrain data there are three methods of animation:

Geometric distortion via vertex animation,
Distortion via morphing,
Distortion on the basis of animated displacement maps.

## Vertex Animation

For the deformation via vertex animation the procedure is as follows: the single points of the terrain to be animated are selected. When the option "ANIMATION" of the respective program is active, for each point distorted after changing the time bar, a new keyframe information is stored.

This method of animation offers the advantage of fast and good results via current manipulations (shifting and rotation).

Furthermore vertex animations can be exported and imported without problems in most 3D programs. Most games or interactive applications are able to "understand" and render vertex animations. Thus, if further editing of the data becomes necessary, vertex animations are a fast and efficient possibility to exchange animated information.

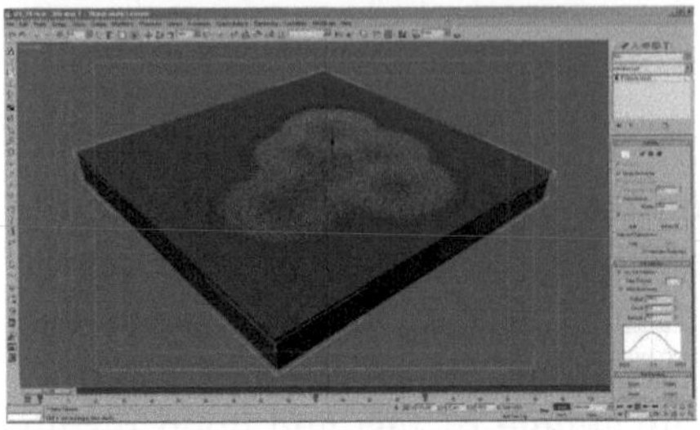

**Fig. 60.** Soft selection of points for the planned animation

**Selection of Points**

The points of the terrain are selected via soft selection.

The time bar is at the frame 0, the option "ANIMATION" is active. This enables a change in time and space for any single point.

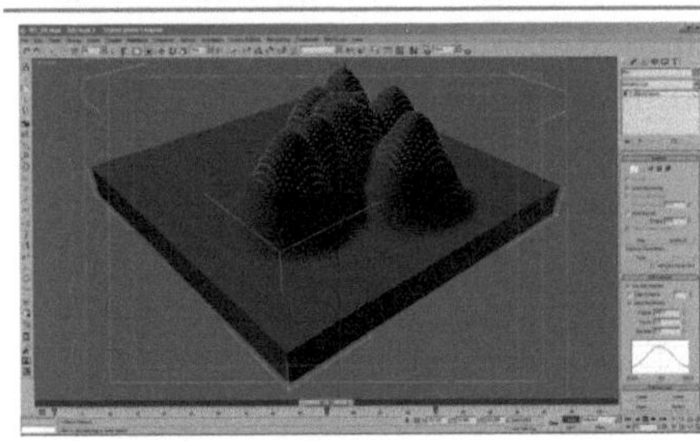

**Fig. 61.** Transformation of selected points at frame 50

**Shifting and Animation**

The time bar shifted to frame 50. The option "ANIMATION" remains active. The selected points are shifted towards Z in the world coordinate system.

When (re)running the animation the interim values of the linear points between frame 0 and frame 50 are interpolated.

## Geometrical Distortion via Morphing

A further method for the animation of terrain data is the use of morphing. In this method which is subject to specific constraints, geometry A is distorted, changed, or "morphed" into geometry B.

The advantage lies in the fact that each stage remains intact and later changes can be taken over into the animation sequence immediately. The disadvantage is the sheer size of the file, because each and every single geometrical stage remains saved in the file. Furthermore, morphing algorithms of different software usually are in-house solutions and are not interchangeable.

The morphing procedure is rather simple:

Usually, a morph object is created, and to this so-called source object are added the various target objects. Then a specific point in time is assigned to every target object at which it will be blended in. The transitions between the single target objects can in most cases be controlled by parameters. However, most morph objects are characterized by the rameters. However, most morph objects are characterized by the constraint that the polygons of the different target objects have to match in number and position.

This means that when doing adjustments via morphing, a new original object has to be created first. Every time this object is copied the respective transformations are done. Even if only one point of the copy is erased, this cannot be used as a target object any more.

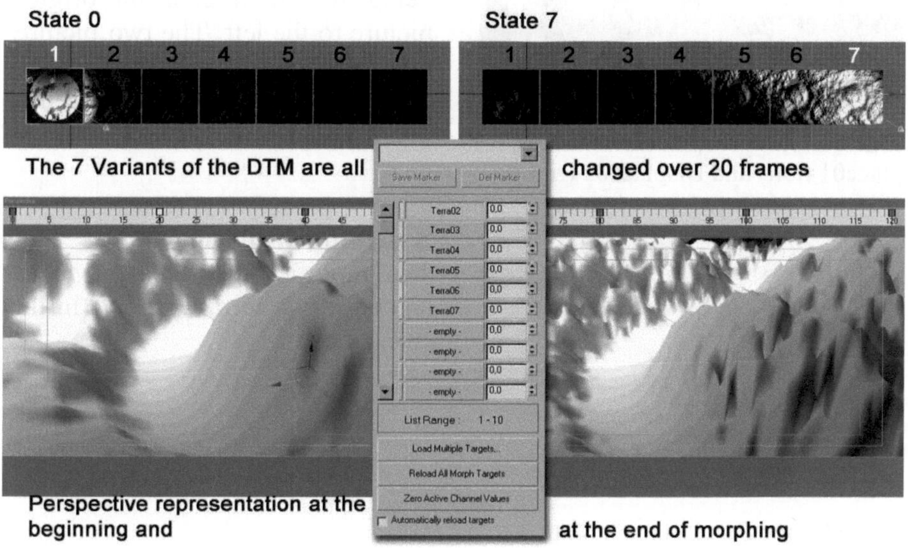


**Fig. 62.** Morphing of a DTM in seven steps

A decisive advantage when morphing remains the easy manipulation of the animation takes and the conservation of the intermediate steps as single objects.

### hint Morphing as Basis of Facial Animation

*In the field of "character animation", the use of morphing techniques via target objects (called morph-targets) is state of the art in games and movie productions.*

## Distortion Based on Animated Displacement Maps

A third and very effective method of animated distortion is the use of animated displacement maps. The procedure is similar to the morphing of geometry, with the difference that here no geometric target objects are created but various displacement maps instead. The animation of these maps then takes care of the respective distortion of the terrain model.

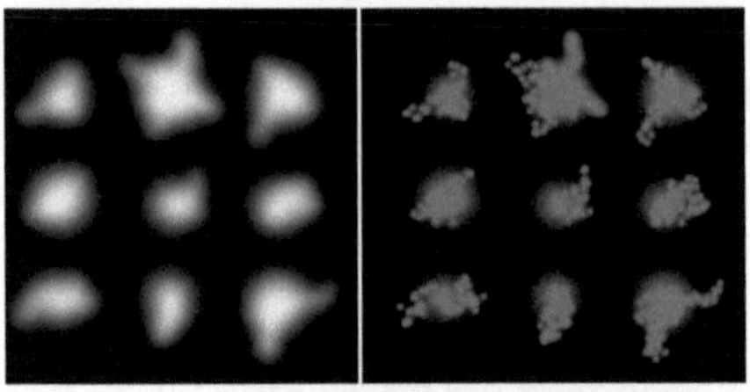

**Fig. 63.** Two Displacement Maps, Displace01 (left) and Displace02 (right)

**Displacement Maps**

Two displacement maps are created in Photoshop, the only difference being the soft filter in the original picture to the left. The two pictures are then saved as GIF-files (because only grayscales are needed).

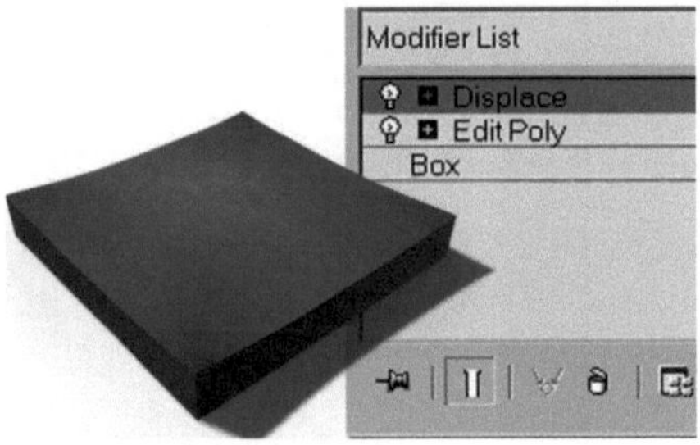


**Fig. 64.** Two Displacement Maps

**Assigning displacement**

The modifier displacement is assigned to the terrain object (in this case a box)

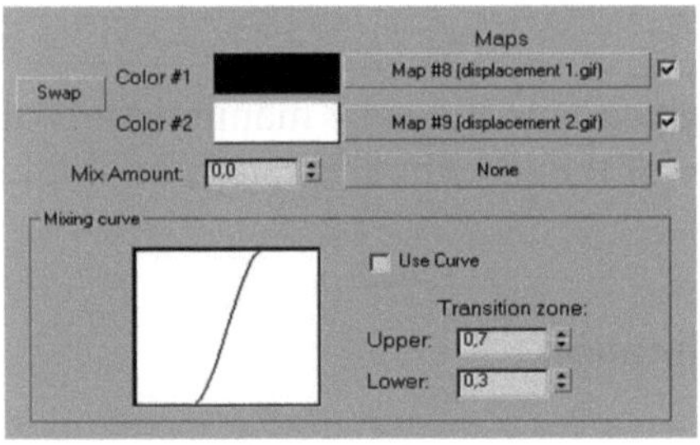


**Fig. 65.** Mix-Map

**Mix-Map**

Similar to the blend material used before, a map is now created consisting of two single pictures, namely Displace 01 and Displace 02.

These two materials can be blended over and into each other. This blend can be animated. At frame 0, the blend is put on 0,0, and the option "ANIMATION" is activated.

At frame 100 the blend is changed to 100. The in-between values are interpolated in a linear way, and the animation is finished.

The advantage is the quick assignment via a map. The animation information is saved in the map, and is thus usable for any other object.

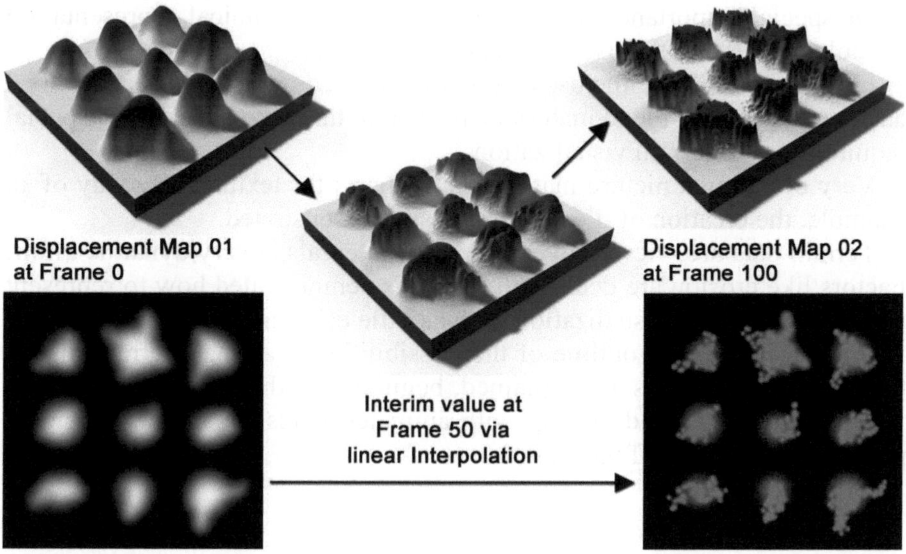


**Fig. 66.** Course of the animation with the help of a mix-map in a displacement modifier

If more than two displacement pictures are needed, instead of a mix-map a movie sequence can be created which contains the blend information as AVI or MOV. In this case, the movie is simply assigned to the displacement-modifier.

## Summary

Digital terrain models can be used for visualization in various ways. The most common methods are by data import of existing terrain models, or by creating one's own optional models.

If a separate modeling environment for terrain models or GIS is not at one's disposal, one has the option of using some data formats for the most common visualizing programs. These are, among others, CAD data, DEM, and Triple data in the ASCII format.

There are various methods for creating one's own (user-defined) terrain models. The most common are the creation via geometric distortion, terrain object (only in 3ds max) and the displacement or 3D shifting method.

The editing or compiling of purely geometrical data only is insufficient. Special attention should be given to the creation of materials. There is a short survey of fundamentals and points dealing specially with terrain.

Of special importance are the possibilities of technical representation via color scales and the use of pixel data as texture.

Further methods are the use of so-called procedure maps instead of pixel data, and the use of blend materials in order to take into account the special requirements of terrain visualization.

Very often, own picture material is the basis for textures. By way of an example, the creation of tileable textures is demonstrated.

Terrain data are, viewed over long time intervals, a very dynamic affair. Factors like erosion are dealt with, and it is demonstrated how to represent those in the field of visualization with as little effort as possible.

Finally, there is an outline of the possibilities to animate terrain distortions. Several methods are explained, beginning with the vertex animation via morph objects, and closing with animated maps for the displacement procedure, to round off the chapter.

# Using the Camera

*This chapter deals with the use of the digital camera. Special attention is given to using the camera as a tool for designing, and to handling the camera for walkthroughs and for special effects.*

To get an insight into the depths of any 3D scene, the obligatory window for 3D application is needed: in general this window is described and defined by a camera.

The virtual camera which serves for viewing and navigation inside 3-dimensional worlds corresponds to and reacts like its physical counterpart. That is why it makes sense to deal with the facts, basics and settings, focus and angle of a "real" camera before attempting to deal with the final issue of a 3D environment, be it a still or an animation.

To be more specific, there are – independently of the tool used for the 3D visualization- some basics, knowledge of which facilitates the "correct" handling of the virtual camera enormously.

Remarks about the nature of projections, a short excursion into the domain of landscape photography, hints for guiding the camera, for designing camera flights and walkthroughs and a few hints and tricks along the way, will be given in this chapter.

## Landscape Photography

3D visualization is modeled on classic landscape photography. It does not make any difference whether in a still or in a movie take, the quality of the result lies in the eye of the beholder,- whether he looks at a digital or real landscape is of no consequence. Thus the following hints refer to landscape photography in general:

Landscape photography means not being pressed for time. Time for planning the shot, for composing the picture and for choosing the right lighting which – usually - is given by natural lighting (daylight).

Excellent general views are achieved by using a wide-angle lens, which can also be used to create an exciting image.

When dealing with details (and clippings) preference is given to telephoto lenses of 100-300 mm.

A panorama picture[1] makes it easy for a user to get the feel of a scene.

Waiting for the right lighting takes first priority, because

The right mood is often found in the morning or in the evening. In both cases the relatively low light of the scene creates a distinct structure which makes the shade an important design element.

During a storm or just before or after sunrise or sunset you can create pictures with a maximum of excitement and suspense.

It is possible to render landscapes equally well in color or in black and white (otherwise, the authors would not have agreed to a black and white print of this book).

Filters are a popular tool for special effects, e.g. filters for a warm and soft color scale, grey scales to darken the sky and polarization-filters to obtain richer and deeper colors.

Having mentioned that the above points are valid for photography in the real world as well as within a 3D environment, one has to admit that the 3D world has definite advantages in some points and is clearly preferable. Lighting can be freely chosen, and even the exposure to the sun can be adapted, if needed. In addition the camera type can be freely chosen which offers quite a few possibilities.

## Camera Type in 3D Programs

In 3D computer graphics two types of cameras are predominant:

Target Cameras
Free Cameras

[1] The term panorama does not mean wide angle – it refers to the actually used film area of the camera. A follow-up research consulting the professionals for panoramas at http:/www.xpan.com is definitely worthwhile.

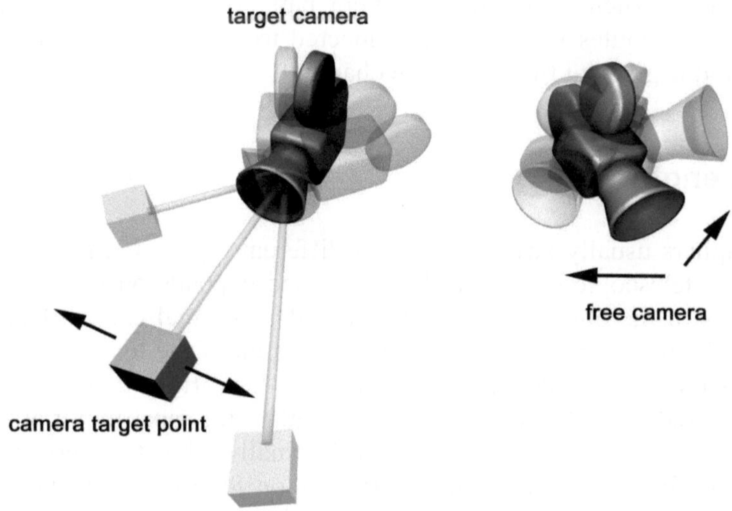


## Target Camera

The target camera is a characteristic feature of 3D visualization. When using a target camera it is possible to define precisely the position of the camera as well as the position of the target point.

Each one of these points can be moved and positioned independently. This has the advantage that the target point, the "**point of interest**" (POI), can be connected to an object, - not only for representation but especially and mainly for animation. The target point can be attached to a moving object, which ensures that the essential point of interest remains in the focus of the camera and in the centre of the point of view.

## Free Camera

A free camera is more like a genuine camera. The free camera indicates the area covered by the camera without explicitly defining the target point. Some programs differentiate between the two types, others offer one type of camera, with the option to set up a specific target point later. The advantage of the target camera is without doubt in the use of stills and in flights over a specific target. To get a target camera to follow a path, for instance, the target point as well as the camera have to be connected to the path, which means a lot more work. For this reason the use of a free camera is recommended when following a camera path, because only the camera with all its attributes needs to be connected to the path. An example for this method is given at the end of the chapter.

## Focal Length

Photographers usually have a choice of different types of lenses. Whether a fish eye, telescopic or standard lens is used depends on the special requirements of the object to be photographed. A normal camera lens comprises different components within the actual unit of the lens (concave and convex parts). The particular arrangement of the various components create the desired optical effect. The focal length is the most important information about the objective of a camera. Normally, objectives cover a specific fixed focal length: 28 mm, 50 mm or 85 mm. However, zoom objectives cover certain ranges of focal lengths: 20 – 28 mm, 28 – 85 mm, 70 – 210 mm.

**Table 2.** Shooting angle dependent on focal length

| Focal length and negative format [2] | |
|---|---|
| Negative format | Standard focal length |
| Small size, 24 x 36 mm | 50 mm |
| Medium size, 60 x 60 mm | 80 mm |
| Large size , 90 x 120 mm | 150 mm |

By the way: the larger the focal size, the smaller the shooting angle! In 3D visualization there is a lens available which can be used for all current focal lengths via parameter control.

The focal length covers the so-called “field of view” (FOW), thus defining the area of a 3D scene to be represented. Each focal length corresponds to a specific shooting or viewing angle. The basis of computer visualization is the small negative format of 24 x 36 mm.

[2] Most affordable digital cameras can not represent a genuine 35 mm camera format (24x36), which means the focal length is increased more or less by a factor 1,5.

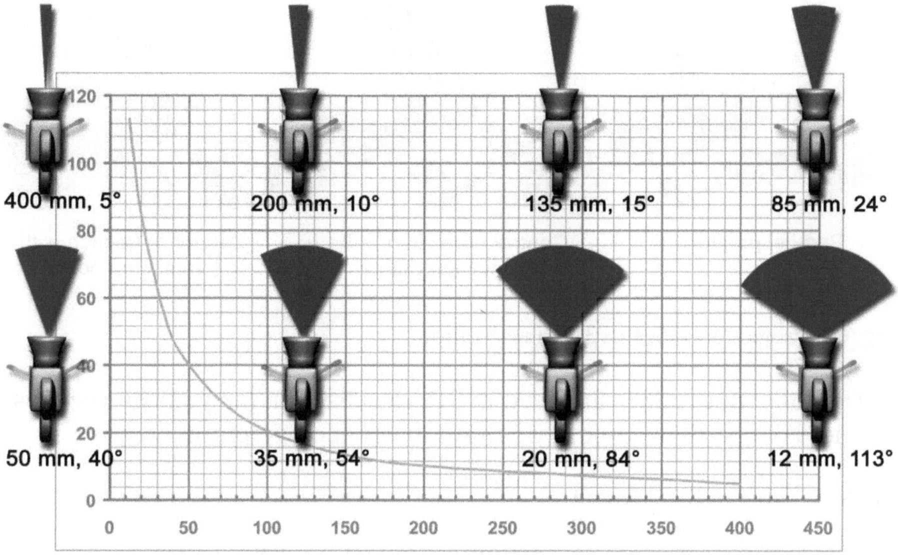


**Fig. 67.** Focal length and shooting angle

The average viewing angle of the human eye lies between 45° and 47°. The standard or normal focal length corresponds approximately to the visual range of the eye (40 mm and 50 mm).

If a longer focal length is used, this is called a telephoto lens. If a shorter focal length is used, this is called a wide angle lens.

A telephoto lens is only able to take a small section of a scene, because it covers only a small spatial angle. But many more details are visible, and the objects are only minimally distorted.

But if the focal length is increased too much, the impression of perspective is lost, and the representation gradually changes into a parallel projection. The following illustration gives a short overview of the projection types.

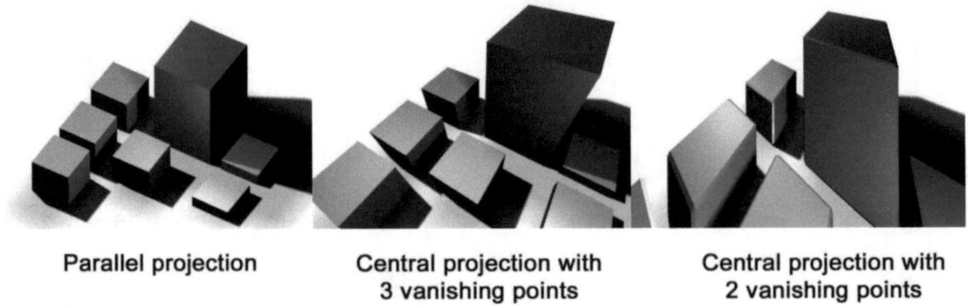


**Fig. 68.** Projection types in 3D visualization

The following values are used in 3D visualization (as you can see these are more or less those used in 35 mm camera photography):

**Table 3.** Table of focal length

| Type of lens | Description | Focal length |
|---|---|---|
| Super wide angle | Short focal length | 10 – 25 mm |
| Wide angle | Short focal length | 25 – 40 mm |
| Standard | Medium focal length | 40 – 60 mm |
| Slight Tele | Long focal length | 60 –100 mm |
| Medium Tele | Long focal length | 100 – 200 mm |
| Tele | Long focal length | 200 –300 mm |
| Super Tele | Long focal length | 300 - ... mm |

Long focal lengths correspond to telephoto lenses, short focal lengths correspond to wide angle shots.

Fig. 69. Standard focal length and "normal" viewing of a scene (50 mm)

### *Standard*

Most cameras in 3D programs have a standard focal length. Within the 30 -50 mm range, this covers nearly everything that is required for a normal landscape visualization.

### *Wide angle*

A wide angle focal length enables the representation of finer details at the expense of perspective. The illustration to the right shows the same scene, but with a distorted perspective.

**Fig. 70.** Wide angle and perspective distortion (20 mm)

**Fig. 71.** Scene drawn up with Tele (135 mm). The fuzziness in the background of the scene was added in a later editing via a hazy filter.

### *Tele*

Focal lengths within the Tele or Super-Tele range offer the possibility of emphasizing scenic details. However, the scene loses its depth.

### *Difference between Tele and Wide-Angle*

The use of Tele setting reduces the visible picture section. The use of Wide-Angle enlarges the visible picture section.

Wide angle emphasizes the distortions with respect to size and gives more depth to the space. Tele settings "flatten" the objects.

**Fig. 72.** Short focal lengths show a considerably larger picture segment than long focal lengths of telephoto lenses; however, this is compensated by a huge distortion in perspective.

## Composition of a Scene

The "right" composition of a 3D scene is subject to the strictest demands of landscape representation. Various factors have to be taken into account, like illumination and lighting, camera angle and picture segment and all of these have to be considered at the same time.

However, the world of 3D animation, as opposed to reality, offers a decisive advantage - there is enough time to play around with alternatives. No external factors will interfere with the necessary settings and tests (that is, apart from pressure from project leaders and impatient clients…).

### Camera Position, Point of View (POV)

No matter how a scene is being planned, it will always be marked by the personal, subjective perception of an individual.

The 3D user decides what is important, which aspect of the model has to become the focus of attention and how this special emphasis is to be represented.

In order to find a successful picture segment with the right perspective, one has to go into the objects from all angles.
The following questions may be helpful criteria, especially when choosing the position of the camera:

Which object is at the centre of attention?[3]
What is the relation of this object to the other objects in the scene, and how can these be most skillfully integrated?
Where is the focus of the subject?
Are there any objects in the scene, which could possibly draw too much interest?
What is the spatial quality of the objects, what are their ranges?
Where is the light source and where are the shadows?
What message do I want to convey with this visualization?

Just try to install the camera according to the above criteria. As a rule, an appropriate position is found and applied by intuition, but at times, it can be advantageous to be able to defend the choice of position by clear reasoning towards a client and towards oneself.

## Position of the camera and placing of the horizon

The positioning of the horizon is a very important factor for the successful composition of a picture. Depending on where the horizon is placed, a picture may appear exciting, calm, well balanced, or at worst, boring.
The position of the horizon is determined by the height and the angle of the camera.
It is recommended not to place the horizon in the centre of the picture. A placing on centre will divide the picture into two halves (have a look at the rules of "golden ratio").

[3] This object is also called POI – Point of Interest

Fig. 73. Horizon in upper third of picture

Positioned in the upper part of the picture, the horizon helps you to imagine the high positioned camera looking at the scene, which gives a good overview.

Fig. 74. Horizon in centre of picture

The horizon is nearly in the centre of the picture, dividing the picture into two halves and causing a quiet, calm mood. However, the calm mood may quickly turn to boredom, and it is recommended to shift the horizon a little into either the upper or the lower third of the picture, depending on the message your are trying to convey.

Fig. 75. Horizon in lower third of picture

The horizon in the lower third of the picture usually corresponds to a low position of the camera, which gives dominance to the objects in the scene, making them appear threatening.

A further, even more extreme way of depicting a scene can be obtained by positioning the camera with respect to the horizon, in other words, there are various perspectives to choose from for depicting a scene, e.g. worm's eye, standard perspective, and bird's eye view.

## Worm's Eye, Standard Perspective, and Bird's Eye View

The worm's eye view is the opposite of the bird's eye view and is preferably used to express the huge dimensions of certain objects, their mightiness, size and strength.

The bird's eye view, as the term indicates, is from high above the scene, which gives the impression of flying over or overlooking the whole scene from an elevation.

The bird's eye view conveys the impression of overlooking the complete scene. This is used especially for the visualization of large construction projects, in order to give a general "overview". It is also very popular in landscape visualization because this offers means and possibilities, which, in real life, could only be provided at extremely high effort and cost.

It is important to use these highly dramatic effects carefully and sparingly. Nothing is as boring as perpetual flights across virtual landscapes.

**Fig. 76.** Worm's eye, standard perspective and bird's eye view

## Picture Sector, Field of View (FOV)

When deciding on the position of the camera, or POV (Point of view), the emphasis is on the type of perspective which gives depth to a scene. When choosing a picture detail, the main emphasis is on the object and its surroundings. Just as the wrong POV, an unsuitable picture detail showing too much or too little can reduce the quality of the image.

**Fig. 77.** Different segments in different formats result in a different focus.

The ratio 4:3 is an accepted format, especially in the size of monitors and in video formats (PAL). When used as a horizontal format it looks very familiar and has the advantage of enabling full scale representations without distortions on nearly all standard monitors (including TV).

A more pronounced horizontal format enables a panorama-like view of the scene which corresponds to the way a human eye sees. However, in this format, with the exception of the cinema, usable areas of the monitor are lost (black strips above and below).

A detail in an upright format can create excitement within a scene and, by emphasizing the vertical aspect, is excellent for high landscape features and plants. However, it is obvious that the upright format may be great for printing, but on the monitor, it is totally useless.

## Format of a Picture Segment

A good test for finding out if a chosen picture segment delivers what is expected is to view the represented landscape and scene in its context.

Very roughly, the layout of a landscape can be classified into plane, hilly, mountainous, green and rich, dry and dusty, and wet and humid.

**Fig. 78.** Extreme horizontal format in 70 mm Panavision with a ratio of 1:2,2

Whereas the first three terms describe the topography of a landscape, the following three categories refer to its attributes and characteristics.

In any case, due to our habitual way of looking at the world with a human point of view, it makes more sense to represent landscapes in a horizontal or panorama format.

One constraint is certainly the method of representation in the field of digital media, where the ratio 4:3 is predominantly used. Even when looking at the cinema or at new forms of TV technology, it is evident that here too the horizontal format will get the upper hand in the long run.

## Dropping Lines

If the camera has not been aligned horizontally, vertical lines will show a tendency to converge above or below.

**Fig. 79.** Dropping lines and how to avoid them by later rectification

Although this effect is more important for classic architecture in building construction than for visualization of landscapes, it should not be neglected. The difference between these two fields of work is often smaller than expected. The bigger the angle of the camera with respect to the level of the picture, the more pronounced the tendency of vertical lines to “drop” will be.

When dealing with object visualizations, like for instance in architecture, try to do without converging lines. Dropping lines do not correspond to our natural observation and should only be used for presenting special effects. You can avoid this exaggerated way of perspective distortion by aligning the camera target point and the camera on the same height.

**Tip Correction of dropping lines**

*Most 3D programs have an built-in correction feature for this effect. As a rule, this is a modifier or effect for the camera used. For this, a quick search in the respective handbook is recommended. In Photoshop CS2 there is a correction tool called aperture corrector. However, if the dropping lines are an important aspect of your visual message, thenby all means use them.*

## Filters and Lens Effects

They are well known from movies and TV, everybody is used to them by now, and although the human eye fails to detect most of these effects in real life, they have become an important part of visualization:

Color filter, grey filter, polarization filter
Lens effects

We have adjusted our viewing habits in such a way that it has become normal for us to come upon these effects, and when viewing pictures, to be astonished at not finding them sometimes. Therefore, it is worthwhile to find out how to create and integrate them into a visualization, even for technical purposes.

In the world of 3D animation, it is difficult to explain the phenomena that one comes across or rather that one has to simulate. The methods applied are very often similar to the procedures in the real world, like for instance lens effects.

A physically caused effect in a real life movie or in genuine photography has to be looked for under render effects in the world of 3D animation. Render effects refer to the editing of a finished picture, and they can normally be added directly in a 3D program or in most picture editing programs as so-called "render post effects". In all the effects mentioned the picture is rendered completely and then edited. In the following, there is a short list of the most important effects, and how they can normally be built into any 3D program:

## Color, Grey, or Pole Filters

This type of filter effect is very simple to create. One method is the use of corresponding colored light sources when illuminating a 3D scene, another variation is the later editing of the complete mood of the color spectrum, in any video post or picture editing program, like for instance After Effects, Combustion, Photoshop, Paint, The Gimp, to name only a few.

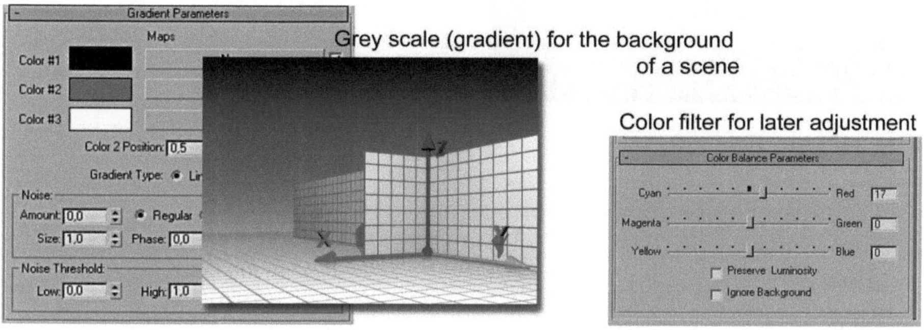


**Fig. 80.** Gradient for the background of the picture and color correction

It is best to create the gradient of grey scales which become more intense towards the top by choosing a suitable background at the time of composing the scene. In nature, the polarization filter helps to avoid annoying reflections and provides deeper colors for the picture. In a 3D visualization, this filter is not actually needed any more. Just like digital color corrections, these effects can be dealt with swiftly during picture editing.

## Lens Effects

Depending on which camera is used, there are two main varieties, namely the lens f-stop or the blending effects and the focusing effects, or rather the lack of focusing.

## *Lens Flare Effects*

**Fig. 81.** Lens flare effects

The most popular lens flare effects are:

Glow
Ring
Star

Not to forget, there are the so-called secondary lens effects. These secondary effects are small rings originating in the source of the lens flare effects and continuing all along an axis relative to the position of the camera. The are caused by the refraction of light on the diverse components of the camera lens. The secondary effects are shifted by changing the position of the camera in relation to the source object.

Fig. 82. Glow effect

**Glow** – this effect creates an aura (the glow) around a light object.

Fig. 83. Ring effect

**Ring** – This effect creates a circular band of colors surrounding the light object.

Fig. 84. Star effect

**Star** – This effect can also be compared to watering eyes squinting and looking into a source of light in cold weather.

## *Depth of Fields – Lack of Focusing in the Background*

A very important aspect, not only concerning lens flare effects, but mainly when creating a scene, is the fact that objects outside a specific reach of the camera appear out of focus. This is not just a characteristic of camera optics but also a means of placing special emphasis on specific objects and areas, which makes this a tool for giving more authenticity to 3D scenes. In photography and movie making, fuzziness is applied either to draw certain elements of a scene into the foreground or to make them recede into the background. Just think of a movie scene where, in a dialog, the two partners are alternately shown in focus (when talking) or out of focus (when not talking).

**Fig. 85.** Varying depth of fields to emphasize the spatial depth of a 3D scene.

Especially when representing a landscape, the option of applying fuzziness to the background is a very essential tool to emphasize the spatial depth of a scene. Like most special effects, this effect is usually a so-called render effect which is only added after the picture has been composed.

## Camera Match or Fitting a Camera into a Background Image

In nearly every area of 3D visualization, placing the virtual camera into a "real" background picture is often required. This makes it possible to combine 3D constructions with a real life environment and to give the impression of realistic planning. The advantage is obvious: one does not have to create a background, and it is easier for the viewer to relate planned projects.

**Fig. 86.** Measuring and recording the surveyor poles

Before taking the picture, measuring aids, in this case, surveyor poles and a tripod are positioned and the area is precisely measured. Additionally, the precise time of day and thus the position of the sun as well as the characteristics of the sky (cloudy, clear) are recorded. The focus setting is noted as well. When all measuring details are known, the background picture is taken.

**Fig. 87.** Integrating the background picture in 3ds max

The background picture is loaded in 3ds max, and a camera is placed. For better orientation so-called camera or tracking points are assigned to the known surveyor poles, a minimum of five tracking points being necessary. Afterwards, the position of the camera in the 3D program is calculated automatically and the camera matches the background image automatically. A precise description of the procedure can be found in the program tutorials of 3ds max. In addition, the function of Matte/Shadow Materials for masking is explained in the chapter "water".

## Guiding the Camera

Giving attention to some minor details will result in a more authentic representation without requiring a lot more effort will add excitement to otherwise boring contents, and will provide a lot more fun during the creation process.

Here the most important points are the movement of the camera through the scene along a previously defined pathway. This can be done either as a camera flight, or as a **Walkthrough**.

## Camera Paths

It is of no consequence whether you install a path to follow the camera directly in AutoCAD, Civil3D, Microstation, Cinema4D, or 3ds max. In principle, the procedure is the same in every case. You draw up any polygon or any spline corresponding to the planned movement of your camera through the 3D scene, connect the camera to this path, adjust the one or two parameters concerning the angle of the camera, define a specific time, and render the result.

In the beginning, the finished movie will cause a lot of enthusiasm, because the work has finally been accomplished, but the second viewing will already dampen the joy. The camera starts abruptly, always moving along at the same speed, passing the most important elements impassively, like a robot. After a short while, boredom will set in and you will shift the time switch in order to get a better look at the prominent spots.

The basic requirement is to get the virtual camera to fly over the created model. Usually the path chosen will follow the given topography, the dominant points of the terrain, or certain planning characteristics intended to be shown.

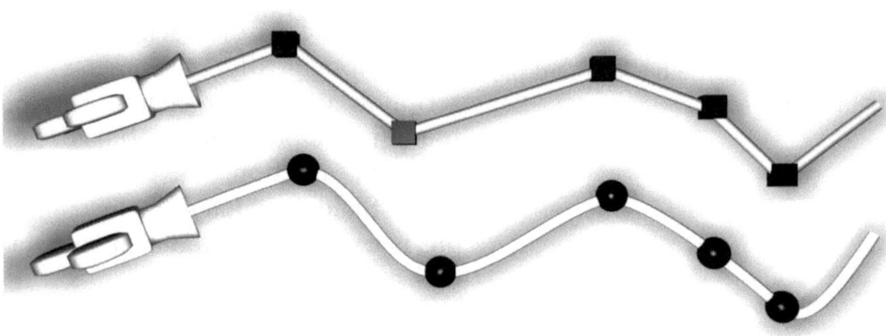

**Fig. 88.** When designing a camera path, one should always give preference to the spline.

It makes perfect sense to use a spline with a "soft" vertex. Bezier points, too, are advantageous. A polygon train with segments following a linear course means a wobbly and shaky camera and abrupt transitions. It should therefore be avoided (except when this effect is intended).

The following example may illustrate such a flythrough scenario:

An arbitrary terrain has been generated. The camera is supposed to give an overview by flying over the terrain. The software applied should be e-quipped with the possibilities of 3D Studio, Cinema, or an equally good tool. The terms may vary according to the visualization software used, but the procedure is similar in all. The path, which the camera is supposed to follow, has already been drawn up.

**Fig. 89.** Example of a scene showing any terrain with a rendered camera path

By answering a few questions before beginning the actual task, you can save a lot of time when editing (excluding the time needed for rendering):

1. How much time is at your disposal for the animated sequence?
2. What is the length and the form of the path?
3. How long does the camera take to fly along the path, and at which speed?
4. Are there any prominent spots, which should be shown in more de-tail?
5. Is it necessary to fly along the whole length of the path in order to get the right impression of the topography?

**Tip Duration of the animation as a basis for calculation**

*Needless to say, the length of the animation is always a good basis for calculating the cost. By splitting up the cost into production expense (data processing, modelling, materials), rendering effort (time per scene and CPU), and editing and post production expense (video post and cut), energy, cost and time become transparent quantities, which gives the customer a much better idea and a lot more confidence.*

## Length of the Animation Sequence

Before attempting any planning, the duration of the animated sequence should be precisely determined, for rendering time is costly.

Furthermore, one should not put the attention of the viewer to the test by never ending flights over virtual landscapes at an even and boring speed. As a rule, the duration defines the possibilities of representation and the speed of the viewing.

The guideline of a flight and the use of a camera path do not imply that the camera has to fly along the whole path. Time can be saved by cutting and zooming in, thus lowering the cost and adding excitement to the animation.

## Length and Form of a Path

The path decided upon should be sufficiently long to reach all important spots. Crossings and repeated approaches to the same spots do not make a lot of sense and may irritate an uninitiated viewer.

One can assume that a path resembling a roller coaster (even though it may look funny) is certainly not suitable for flying over the planned renaturation of a running stream. The form of the path should follow the given topography. Like a walk, the path is an aid for exploration and should be treated as such.

### Tip How to calculate frames for the length of a path

*Assuming the length of a path is 35 m, the average walking speed is 1,5 m/s, and the result is to be rendered for a NTSC resolution. The following number of frames will be needed:*

***35 m/ 1,5 m/s = 23 s, 23 s x 30 fps = 690 frames!*** *And for PAL:*
***35 m / 1,5 m/s = 23 s, 23 s x 25 fps = 575 frames!***

## Duration of Flight

It is common sense to keep in mind a few guide numbers: a helicopter will most certainly not fly faster than 200 km/h maximum ( more like 150 km/h or 46,7 m/s) and a pedestrian walking around and looking at the place (walkthrough) will certainly not move faster than 5 km/h (~ 1,5 m/s).

Taking into account these guide numbers, a realistic speed for viewing is guaranteed.

After calculating the maximum running length of an animation (normally given in minutes and seconds), and taking into account the available time and the assumed speed, you quickly arrive at a rough guide number for the maximum covered distance.

It is always a good idea to draw up a preview of a scene by using simplified geometry. These short movie sequences are called animatics, and they can make life considerably easier because they facilitate reaching a decision about an animation, before tackling the often time consuming renderings of a complete landscape scenery.

## *Time Variation*

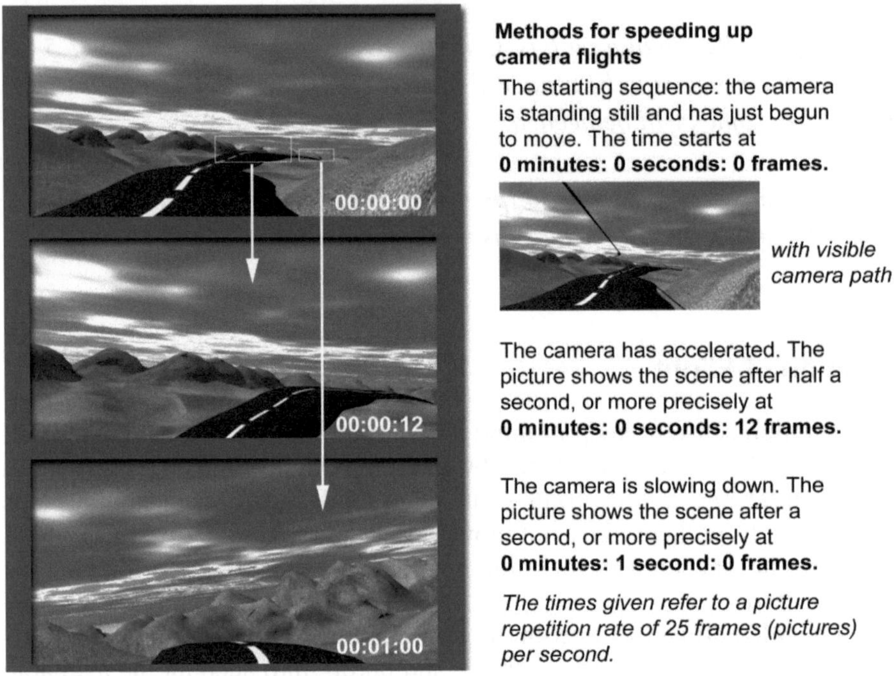


**Fig. 90.** Methods for speeding up camera flights

It is not always necessary to show the entire path of movement. In some cases, pieces can be cut out without spoiling the content of the animation.

Just imagine you want to present a landscape planning. The important part is the design of the hills at a distance of about 1,5 km. In order to get to the information placed in the foreground, one has to fly along the road leading there. The entire duration of your animation may not exceed 1,5 minutes. A walkthrough in the area of the important terrain segment may not exceed the abovementioned pedestrian speed of 1,4 m/s.

Flying over the 1,5 km stretch of road alone takes 36 seconds, which is a third of the entire time at your disposal. Much too much.

The road is important and therefore cannot be ignored. However, there are methods of speeding up the process.

The first method is a simple cut. You let the camera fly along the road towards the “important” area, the hill. The sequence starts with a camera still, the camera accelerates for about 2 seconds – cut – the camera has just arrived in front of the hilly area and is already slowing down.

A large chunk of the road may be missing, but the viewer will automatically replace the missing piece of the movement in his mind. Thus, you have gained a lot of time for the important sequence.

A further option would be to accelerate tremendously over the short stretch of road and then brake sharply, so that the camera is actually moving along the entire distance. Nevertheless, this movement appears only for a short time, does not seem to be dominant, and does not deflect the attention from the important contents.

**Tip How to project a camera path onto the landscape**

*Sometimes the camera path has to follow the contour of the landscape. There is a method for shifting the supporting points of the path manually (normally a spline). For this method Itoo Soft has written a very nice Plugin for Max. The Plugin is called Glue and can be downloaded free of charge from the homepage www.itoosoft.com.*

### *Prominent Points (Landmarks)*

In nearly every 3D program, there is a method of connecting an object directly to a path. Therefore, this appears to be easy work. However, as a rule there is more to it. As will have become clear by now, it makes sense to play around with the speed. Acceleration and slowing down can add excitement in some places.

Prominent features, sometimes called landmarks like for instance a building, a group of plants or the special constellation of earthworks should not be passed at the same speed as the rest of the scene.

Just before the sequence of the road was shortened by acceleration or cutting, it makes perfect sense to lengthen a sequence by slowing down the camera near prominent features, to change the direction of the movement a little, and, if appropriate, even to vary the aperture.

The camera could slow down, for instance, when approaching the POI, then come to a near halt and change the view by zooming in.

### *Controlling the Camera*

It would be very expensive to execute all of the steps mentioned above with a camera connected to a path. Especially when certain parameters have to be changed afterwards. Here, the use of animation aids can offer exciting possibilities. These construction helpers (also called dummies) enable a more finely tuned steering and some additional freedom of control. This could be described as follows:

### *The Camera follows an Object along the Path*

This setting is extremely well suited for single sequences where the camera follows a specific object. A variation could be the camera flight along the object, another variation is a moving object.

Figure 89 shows a possible setting for a dummy connected to a path. The target point of the camera is connected to the dummy. This ensures that the camera is always directed towards the dummy. In the example, the dummy as well as the path has been superimposed by way of illustration for better understanding.

In Figure 90 the camera has been connected to a second dummy and the settings for the camera target remain the same. The advantage in this case is that the camera, too, follows the path. This type of animation is already very close to an optimal steering of the camera along a camera path. The hierarchical connection of the objects in the order: path of motion – dummy – camera target point (or camera) – ensures that every object has to follow its predecessor, but can itself be moved freely. This means that, the camera, although connected to the movements of its dummy, can, for instance, at any time be freely shifted or rotated.

When adding yet another dummy, as in Figure 91, this has the advantage that ALL the information with respect to the animation is connected to dummies. This means that, the camera stays free of any direct animation information (keyframe animations).

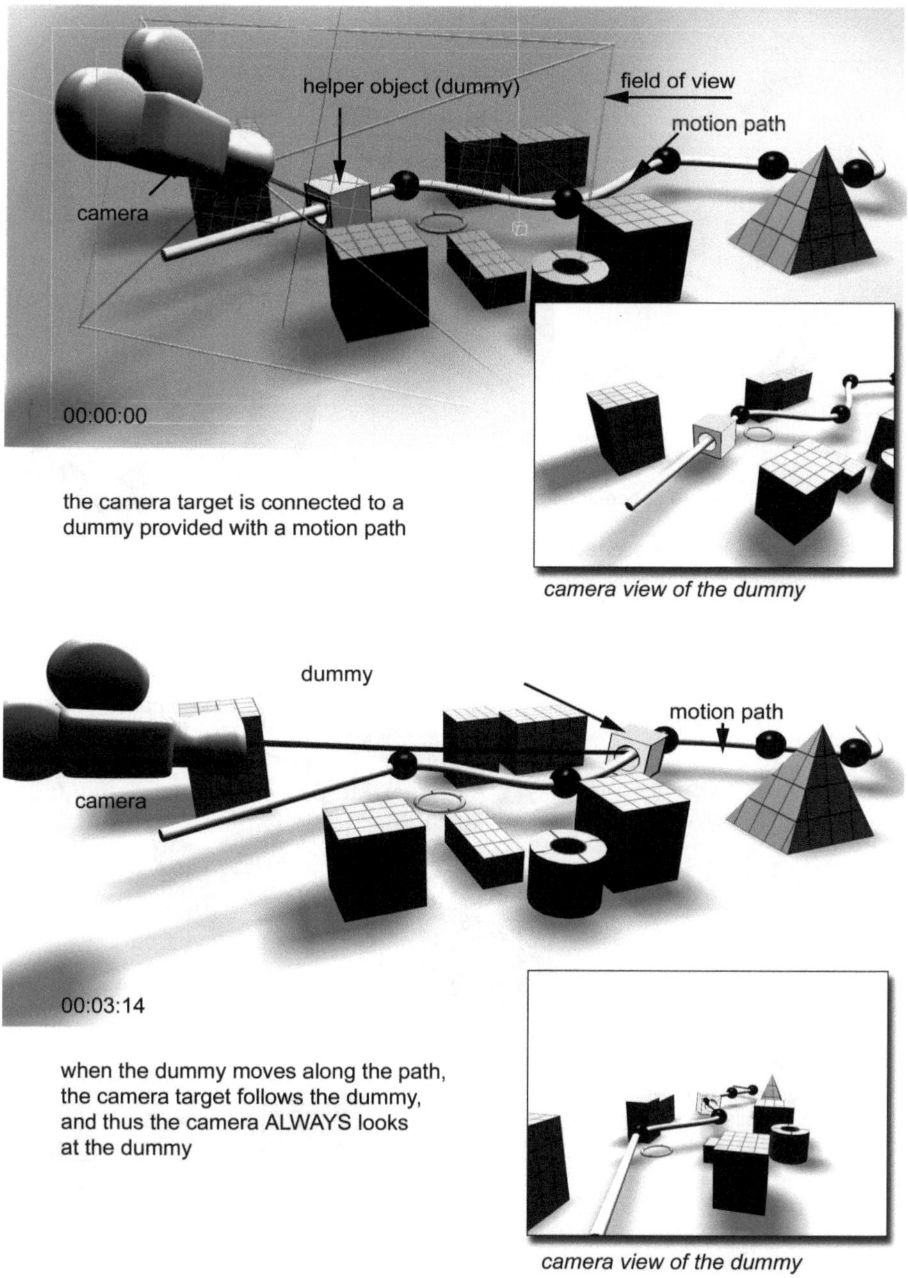


**Fig. 91.** The camera target is connected to a dummy provided with a motion path

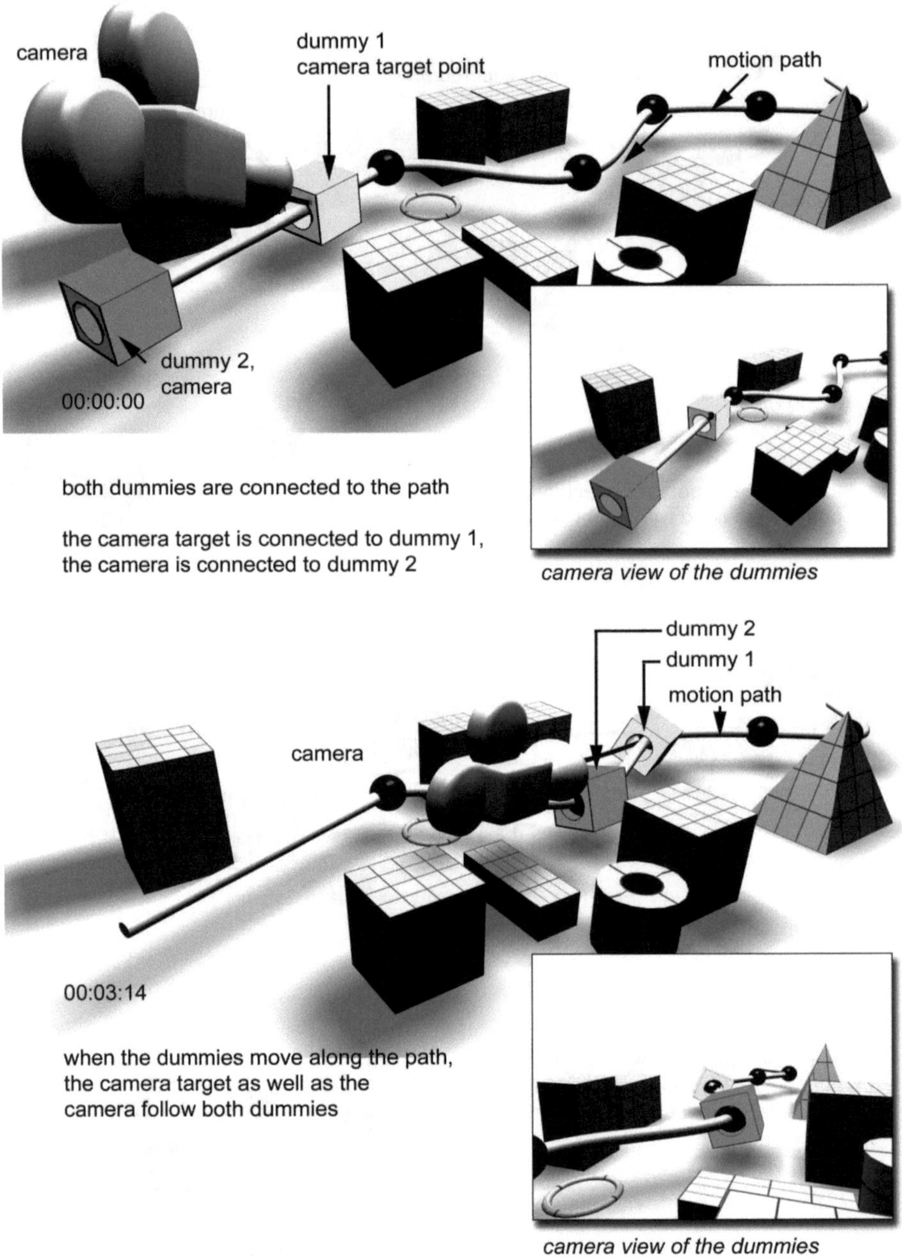


**Fig. 92.** The camera target point is connected to a dummy provided with a motion path, the camera follows the second dummy

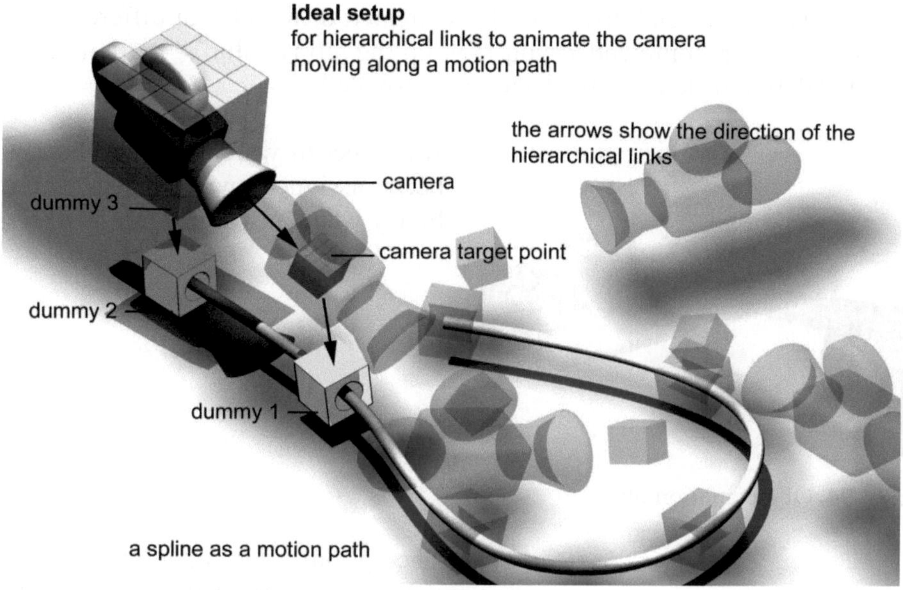


**Fig. 93.** The ideal setup, where the camera itself contains no animation data (keyframes) but is animated only via links.

## Motion Blur

Motion blur is an important aspect for adding authenticity to a representation.

In a "real" camera, the aperture is open only for a specific time. Very fast movements during this time slot will cause blurriness in the picture, or rather in the pictures of the film.

From the point of view of a standing camera, this means that fast moving objects become blurry, and from the point of view of a moving camera, objects outside the field of view of the camera become blurry.

In 3D animation, motion blur can usually be achieved in two ways:

1. The motion blur is directly assigned to a moving object. Consequently, as soon as the object moves, the blurriness is added as a special effect when rendering.
2. Depth of field is assigned as a special effect to the camera. Everything outside a specific distance from the focal area is blurry. The blurriness increases with increasing distance to the camera.

Because the special effect of blur is treated differently in different programs, it is recommended you look up the terms "blur", "motion blur", and "depth of field" in the handbook provided.

Fig. 94. Object motion blur

An object to which motion blur has been assigned will appear more blurry when moving faster.

Fig. 95. Blurriness outside the objects on which the camera is focusing

The special effect of motion blur is assigned to the camera. Similar to depth of field, all objects outside a defined area become blurrier with increasing distance to the camera.

## Summary

This chapter dealt with the similarities and differences between the real camera and the camera in the world of 3D animation.

After looking at some basics of landscape photography, our attention returned to the camera and its use in a 3D program.

In 3D animation – as opposed to the real world – there are two types of cameras, namely the free camera and the target camera. The first corresponds to the "real" camera, the second is a typical representative of the 3D world.

You have become familiar with the focal length, and know by now that a large angle of the focal length (28 mm) means a wide angle, and that a very small angle (105 mm) designates a tele-objective.

When putting together 3D scenes it is important to place the point of view, the POV or in other words the position of the camera in such a way that the important structures are shown as well as possible.

In this respect, the position of the horizon and the various perspectives are an important part of the design. There is a distinction between worm's eye normal and bird's eye perspectives.

The picture segment, or field of view (FOV), helps to emphasize the desired mood.

Converging (dropping) lines are a phenomenon not only in architecture, they are also a nuisance in landscape representation. A method for later removal was introduced.

You are now familiar with some of the typical lens effects of real lenses and their simulation in 3D design and you have learned about some basic characteristics of the moving camera. Important aspects in this area are especially the animation of a camera along a path, and the aspect blur, as shown in motion blur and depth of field.

A lot has not been mentioned, and some aspects have only been lightly touched on, but the rough sketch of the camera in general and in digital simulation in particular will have removed some uncertainties.

Those readers who would like to know more about constructing a scene and designing and handling a camera, will find some suggestions for further and more in-depth literature in the appendix.

# Lighting

*Where there is no light, there is no shadow.... And shadows are absolutely essential for a successful visualization ..*

## Introduction

In landscape representation there is only one significant source of light, and that is the sun. Without the sun there is no light, not even moonlight. This one source of light, mixed with a couple of atmospheric effects, is all that is needed to generate an incredible range of light variations.

Today, most 3D programs include rather sophisticated methods of simulating sunlight. The position of the sun, the geographical location on the globe, the state of the sky (cloudy or clear), the time of day, and many more details can be assigned. In most cases the results are quite convincing; the user merely has to find the respective menu.

**Fig. 96.** The sun provides light and shadow

But sometimes this is not as simple as that. These "ready made systems" can quickly be drawn up, but very often, complicated calculations like GI

(Global Illumination, e.g. radiosity) are needed, which, although they may seem enticing due to a fantastic result, unfortunately still require many resources (memory and processor capacity).

So, it is a good idea to go into a little more detail with respect to the subject of illumination, in order to become familiar with the standard sources of light in 3D programs. The correct illumination is a further very important aspect of designing a scene.

**Tip Lighting and storyboard**

*It is advisable in any case to write down the intended settings for illumination in a script or storyboard, and to consider carefully where exactly the light sources are to be positioned. The light source together with its respective shadow gives a direction of light to the picture and is thus a very important component of design.*

Especially when integrating a 3D scene into an existing background picture, it is extremely important to adjust the lighting of the scene to be integrated to the color and direction of the lighting of the background picture.

Basically, the various light sources can be classified according to the type of light and to their function.

Looking at the various forms of illumination and how these are simulated in the world of 3D programs will shed some light on the subject.

## Types of Light

Every 3D program has its own way of dealing with the terms and the forms of light sources. Different forms of light sources are provided as well as different terms. In most cases it is difficult to establish later, in other words after the picture has been rendered, which type of light was used. Furthermore, the use of a specific light source is influenced by personal taste, preferences, likes and dislikes.

There are users who present fantastic results with only a few spotlights, whereas others prefer the use of sunlight systems based on complex radiosity simulation systems.

Most of the 3D programs have certain basic kinds of light in common.

These can be described as follows:

Point light or omni light
Target (direction) light or spotlight
Directed light or parallel light
Area light

**Table 4.** Kinds of Lights

| Point Light | Target Light | Direct Light | Area Light |
|---|---|---|---|
|  |  |  |  |

## Point Light or Omni Light

A source that sheds light evenly in all directions (omnidirectional) is called point light or omni light.

The classic example for this light source is the light bulb.

By the way, the biggest known point light is the sun. However, in a computer simulation, the sun would never be simulated as a point light, because of the sheer size of this particular point light. Due to the real, immense distance of the sun to the earth, the sunrays hit our blue planet in a nearly parallel way. Thus one would rather choose a parallel light.

Every method of calculating the shadow of a light source requires memory space and capacity. A point light will always require more calculations than for instance spotlights or parallel light.

In any case, point lights are not necessarily first choice for (merely) illuminating a scene. For the main or directed lighting, the use of a spotlight or of a parallel light is recommended, whereas point lights are an excellent means of emphasizing the mood of a scene by adding lights in specific areas.

**Fig. 97.** A point light sheds its light evenly in all directions

## Target Light or Target Spotlight

Because of the cone-like spreading of its rays, a spotlight is actually first choice when simulating a floodlight, a flashlight, or other artificial light sources. One can assume that in classic computer graphics one or more spotlights are usually the main light sources for illuminating the represented scene. A spotlight makes it possible to illuminate specific objects, as opposed to a point light.

Similarly to a target camera, most 3D programs include the option of linking the target point of a spotlight to a specific object, ensuring that the desired object is optimally illuminated at all times.

Furthermore, of all light sources, spotlights are generally those, which offer the most configuration settings.

**Fig. 98.** A spotlight sheds its light in one direction only, in the form of a cone, like a torch

## Direct Light or Parallel-Light

When thinking of the sun as a source of point light and at the same time taking into consideration its distance to the earth and the small size of our native planet, it quickly becomes evident that the rays of light originating from the sun and hitting the earth are rather more parallel than cone-like. Leveled light is also called direct light, distance light or simply sunlight.

The illumination vector is the same for all rays of light, which implies that the shadows of all objects point in the same direction.

As far as the application of target light is concerned, people's tastes differ. Some say that parallel lighting is the best light source for the environment; others prefer it for simulating sunlight.

**Fig. 99.** The rays of a direct light pass along parallel lines

## Area Light

While the source of a point light is a minute point, the area light is defined by a specific "size". In a spherical light source, the size refers to a diameter, and in a rectangular light source, it refers to a length and a width.

In this way, when using area lights, the rays hit the object in a parallel fashion, just as direct light! Therefore, if an area light is very small it acts like a point light, creating similarly sharp contour-like shadows. With increasing size of the area light the contours of the shadows become softer, and the overall illumination turns somewhat dimmer. It should be mentioned, however, that area lights require huge resources. Therefore, the use of area lights should rather be avoided in landscape representation, even though the effects may appear very realistic.

How the different types of lighting can best be used for outside illumination, i.e. in the visualization of landscapes, is described in more detail in the example "daylight with standard light sources".

**Fig. 100.** The area light spreads to the whole of an area and causes soft shadow contours

## Light and its Definition according to Function

Apart from the different kinds, there is another way of looking at light, namely by describing its function. This method appears to make a lot more sense, because in the end it is the personal likes and dislikes of the user which decide about what type of lighting should best be used in a specific case.
Classified according to function, there is

Ambient Light
Main Light
Fill Light

## Ambient Light

The ambient light creates the basic mood and provides the basic brightness as well as the basic coloring of the 3D scene.

Thus the ambient light is responsible for adding a particular mood to the illumination of a scene. For creating an illumination mood at night, predominantly dark colors in shades of blue are used, whereas in a daylight illumination light shades of yellow dominate, right up to a blinding white for simulating sunlight at midday.

**Fig. 101.** Illumination of a scene by area light only. Although a certain spatial quality is suspected, the scene is completely lacking in shadows and thus appears very flat.

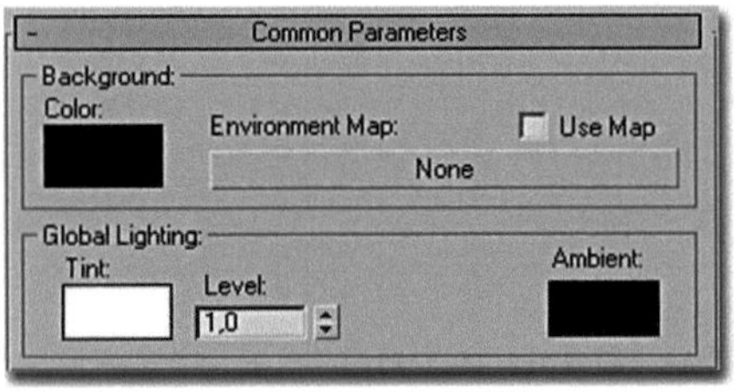


**Fig. 102.** „Global Lighting" – environment light in 3ds max

**hint Ambient Light**

*In most 3D programs, this type of lighting is defined via a scale of color and intensity. Environment light does not normally provide shadows.*

In some calculation procedures, which do not attempt a physical light simulation, the environment or global lighting simulates the diffuse distribution. In other words the illumination of single objects is achieved by the reflection of other objects.

However, as exciting as this may seem, one should not forget that a complete and even illumination as generated by global lighting will result in a very unrealistic light composition, because this effect does not exist in real life.

When a 3D scene has been illuminated only with a global light, it will appear unrealistic, and the viewer will immediately recognize an inconsistency, without actually being able to say why.

The environment light will be adequate for the initial composition of a scene, but it should never be used for the final illumination.

## Main Light, Key Light, Guiding Light

The main, key or guide light is the relevant source of light in a scene. This type of light defines the direction of the light and thus the direction of the shadows.

In order to integrate the design elements or components of shade distribution, the main light is often defined as a source of light illuminating the scene from above and from an angle. The main light is normally the brightest source of light used in a scene.

**Fig. 103.** The main light in a scene generates shadows and controls the direction of the light

### hint Sunlight systems

*Nowadays, most 3D systems provide rather sophisticated sunlight systems. These do not only include the sun as the source of light, and its course across the sky by day, but also the component for the respective diffusion.*

The main light, as it is the most significant source of light and determines where there is light and where there is shadow.

Additionally used sources of light serve to fine-tune the characteristics of the main light. These additional lights are usually called fill lights.

## Backlight

In landscape visualization, the backlight is used for softening the contours, especially in daylight illumination. Instead of using the term for a light source exactly positioned opposite the camera, in this context it refers a light source positioned opposite the main light source. The backlight should not have an intensity exceeding 40% of the main light source, and shadows should be strictly avoided in a backlight.

## Fill Light

The fill light lightens up the shadows caused by the main light, helping to achieve softer borderlines of the shadows.
The brightness of the fill light should be lower than that of the main light.

**Fig. 104.** By adding filling lights, the shaded areas caused by the main light are somewhat lightened up

### hint Point Light as Fill Light

*Normally point lights (omni lights) are used as filling lights.*

Apart from the illumination by sunlight only, there are mixed installations where construction works or streetlights are used as main artificial sources of light. In this case, the light sources are mostly spotlights or parallel lights. Incidentally, spotlights or parallel lights are the most frequently used sources of light.

These four main sources of light, as described above, are sufficient for providing a scene with a basic illumination.

However, to attain at a really satisfying illumination, it makes sense to study the simulation of the natural source of light, the sun and daylight, in more depth.

### hint One source of light is seldom enough

*Do not try to save on light sources. Only one source of light is usually not enough for illuminating a scene, not even in light simulations like Radiosity or Global Illumination.*

When looking at the subject in more detail, you will quickly come upon terms like reflection, diffuse distribution, indirect lighting, light color and so on.

One or two sources of light are not all there is to illumination. It is an enormously complex interplay of different wavelengths, surfaces and reflection characteristics, and is certainly one of the most extensive subjects of computer visualization.

## Illumination Procedures – a few introductory Remarks

The development of illumination procedures is proceeding very quickly. During the last couple of years, special attention was given to illumination models. There still is an unfailing strive for perfection, for representing nature in the most perfect way possible. Global Illumination and Radiosity are two examples of this endeavor.

Think carefully about whether or when to work with which illumination method. The appropriate illumination will depend on the method of modeling, the time you are prepared to invest (calculation of computing time), and of course on the demands of the client.

The decision should also take into account your experience. Do not make the mistake of starting with a new method while being pressed for time to finish the project.

Most 3D programs include several tools for implementing an effective illumination that turns out well.

For any illumination it is essential to first answer the following questions:

Which light for which purpose?
How exactly do light and shadow relate to each other?
Which kinds of shadow, Raytrace or Shadow-Map, will be implemented?
Which illumination systems are available?
How do you achieve a satisfying exterior illumination?

## Lighting Methods

There are a great variety of methods that can be employed for lighting 3D scenes, some of them requiring huge calculations. Basically, there are two methods of lighting: the method of simple illumination with standard light sources based on simplified physical parameters, and the method of a nearly real-life simulation.

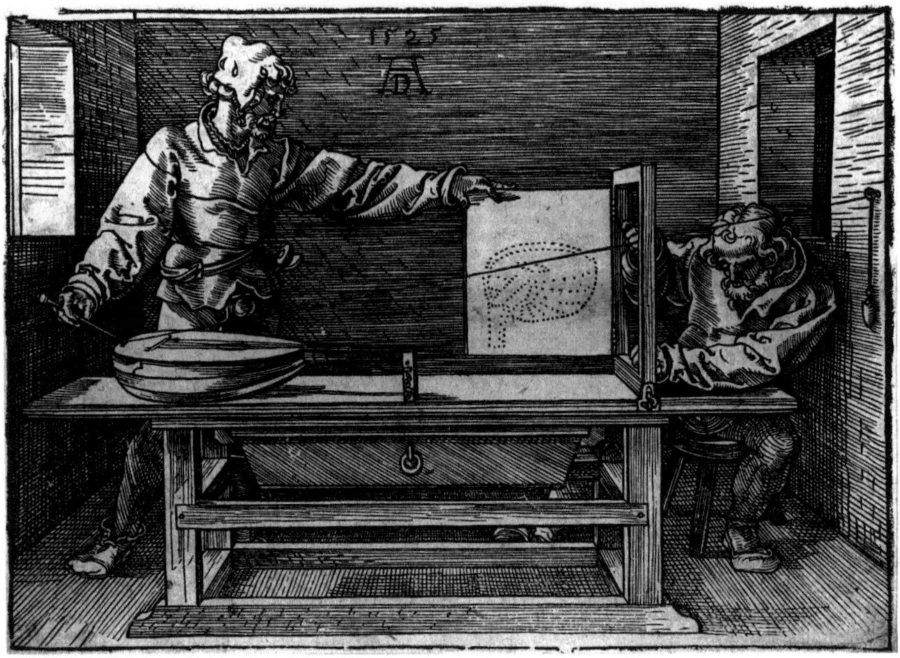

**Fig. 105.** The copperplate engraving "Teaching how to measure" by Albrecht Dürer clearly demonstrates that the subject of tracing sunrays and their representation on projection levels is not a modern invention.[1]

The first method addresses the rather simplified methods of standard light sources of most 3D programs, refraining from the actual physical simulation. Here, experience, a good instinct, and a precise observation are top priority. If these requirements are met, high quality results will be achieved in a minimum of time.

The second method is based more on the physical reconstruction of complex procedures, relying less on instinct than on mathematics, which is the basis for representing nature as near to real-life as possible. Unfortunately, this kind of simulation is in most cases extremely calculation and time intensive.

Both methods have advantages and disadvantages, but both are important in the world of 3D modeling.

The following examples will briefly explain how the methods for standard light sources and for photometric (i.e. physically nearly correct) light sources work.

[1] By courtesy of : Albertina Reproduction Department, Albertina Place 1, A – 1010 Vienna

A few important terms when dealing with this subject are:

Local Illumination
Global Illumination
Raytracing
Radiosity

It has certainly become obvious that illumination methods are in fact rendering methods. Rendering procedures, in other words the final picture calculation, are defined by the method of calculating the illumination of a scene.

## Local Illumination - LI

As the term implies, "Local Illumination" describes how the surfaces of specified objects react to the action of light when illuminated, or more precisely how they react with respect to reflection and absorption. Regarding the rendering procedure of a 3D program, this implies that calculations are done for each object separately.

An interaction, or rather an indirect illumination between the objects, does not take place.

Calculating a scene only by means of a LI method makes the scene appear flat and very unrealistic in most cases. Even when shadows are added, the component of diffuse reflection is completely missing.

Classic examples of LI methods are Flat Shading, Gouraud Shading, and Phong Shading.

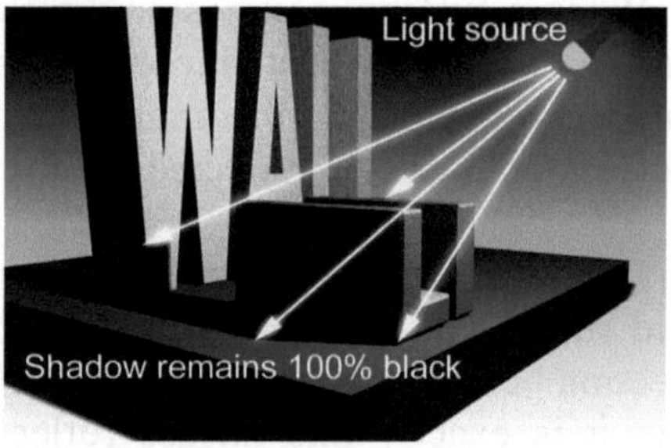


**Fig. 106.** Local illumination with one source of light

The behavior of a surface with respect to its illumination characteristics and its reflection is defined via the material. This method is called Shader. For visualizing landscapes, taking LI as basis is certainly the quickest way to obtain acceptable results within a very short time.

In a LI application, the color of a specified point on a surface within the 3D scene is calculated as a function of the illumination in the actual model. There are no further calculations.

## Global Illumination - GI

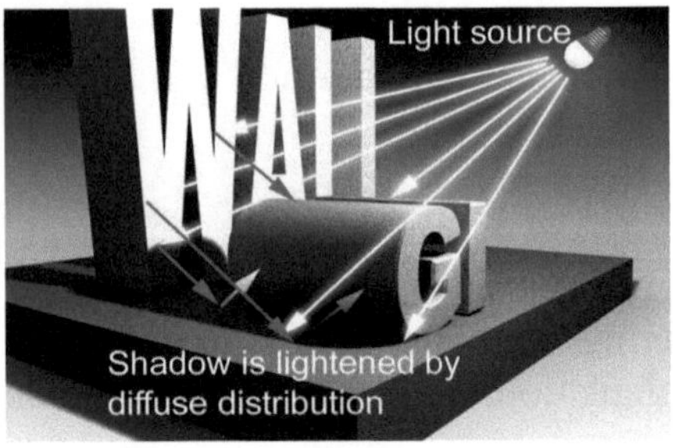


**Fig. 107.** Global Illumination with one light source

Calculation methods, which take into account that the light is reflected by the objects, thus creating diffuse lighting for the other objects, are called Global Illumination methods.
Two examples for GI are: Raytracing and Radiosity. The light source in figure 107 generates energy, so-called photons. These are radiated by the bulb evenly in all directions.

When particles of light hit a surface, some parts of the energy are absorbed, while one part is reflected, this reflected part having been colored by the surface hit last. When this part reaches another surface, this surface will be partly lightened.

When a surface is very shiny, reflecting the light like a mirror, the light hitting it will be reflected more or less according to the rule: "The angle of incidence is equal to the angle of reflection". The rougher, or soft, a surface is, the more the reflected light will be distributed.

In landscape representation the only reflecting surfaces are water, ice, and snow. The remaining parts of the landscape show a more diffuse reflection behavior, distributing the reflected light more or less widely.
But it is exactly this diffuse reflection that is responsible for attractive landscape visualization.
Imagine the mood in a forest at noon, the light hitting the forest floor unfiltered and in brilliant white. This scene may be very well illuminated, but the viewer expects a diffuse lighting caused by the green leaves. The discrepancy is obvious, even if it can not be traced to a specific detail, the scene appears unconvincing.
When compared to LI, Global Illumination certainly is nearer to reality, but this method requires an incredible amount of time, depending on the procedure used for the calculation of the picture.

## Raytracing

Raytracing is one of the very first methods developed for GI. Raytracing is in a way the movement from the eye of the viewer, or rather, from the camera, right into the scene until the ray has arrived at its point of origin, the light source. This procedure is followed for every pixel on the picture that is to be rendered. It is quickly evident why a higher resolution means longer calculation times.

Raytracing is an excellent method to compute direct lighting, shadows, reflections, and reflecting surfaces. Even refraction can be very precisely calculated via Raytracing. However, Raytracing methods have their price, and that is patience.

As a procedure method of GI, Raytracing takes more effort than, for instance, a LI procedure via Scanline-Rendering.

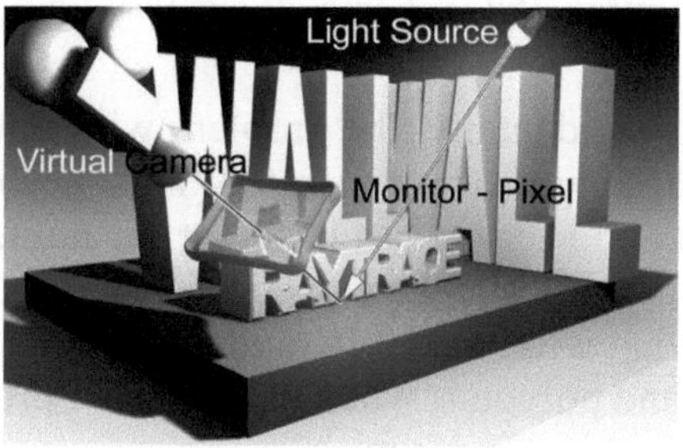


**Fig. 108.** Raytracing – the camera follows the ray of light across the screen and through one pixel (e.g. 1280 pixel width x 1024 height) until it hits an object and is reflected towards the light source.

Furthermore, a mere Raytracing procedure does not take into account the diffuse reflection, which is so very important in landscape visualization.

## Radiosity

In a method like radiosity, the picture calculation of light distribution within a scene is done according to physically correct guidelines. The simulation of light is not, as in the previously mentioned procedures calculated on the basis of a simplified distribution of rays, but works according to the rule of the preservation of energy. This means that a light source contains certain parameters like luminous flux (lumen lm – light energy per unit of time), illumination (lux lx - lumen per square meter), light intensity (candela cd, light intensity of a wax candle), and luminance (candela per square meter, one part of the light reflected from a surface).

Furthermore the application of such a procedure requires scenes with true, real-life dimensions. A very important aspect of radiosity is the fact that the decrease in light is taken into account. This is not automatically done by most standard light sources, and if so, it is hardly ever correct.

### hint Light Decrease

*With increasing distance to its origin, the intensity of light decreases. Objects which are nearer to the light source appear lighter than objects which are far away from it. This decrease in light is described as a square function. This means that the light intensity decreases proportionaly to the square of the distance to the light source. The presence of clouds, fog, and similar atmospheric effects will amplify the decrease of light.*

Incidentally, the problem with radiosity procedures is that an additional mesh has to be generated. This mesh is drawn over on the existing geometry and serves as a basis for the numerical equations. This additional calculation mesh takes up additional resources, and it is not difficult to imagine what will happen to landscapes with many details and some hundred thousand polygons when using such a procedure; in most cases, for a very, very long time, nothing.

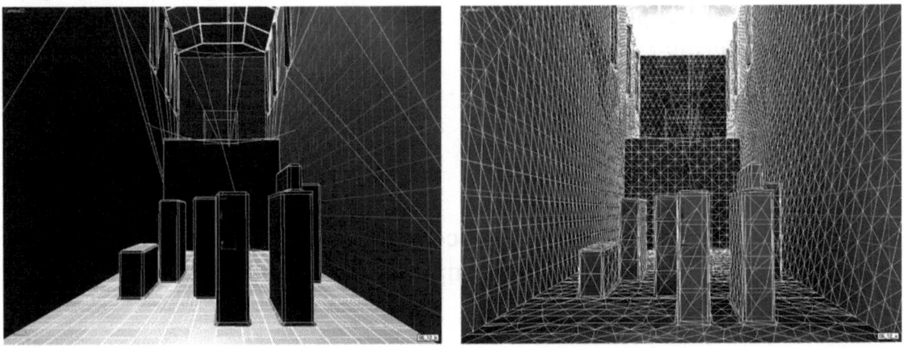

**Fig. 109.** The left picture shows screenshots of a scene before – and the right picture after – the calculation of a radiosity procedure. One can see the numerical mesh.

So many methods and procedures – but what is the best way to get satisfying results? There is no magic formula, but there are some hints and recommendations.

Especially for landscape visualization, which is blessed with many polygons and requires very long rendering times, it is worth the while to take a closer look at standard light sources. By investing a little extra time, quite a convincing illumination can be put into practice.

## Simulating Daylight using Standard Light Sources

Daylight means sunlight; independently of the calculation methods, sunlight can be created by different methods.

As in the procedures previously dealt with, it is important to understand which components are relevant when creating the illumination.

Although the sun is more like a point light (see previously introduced light types), it is situated at such an enormous distance from the earth (and considerably larger, too), that the rays of light finally arriving here seem to be parallel.

A further important aspect when taking pictures outside is the diffuse distribution of light by the sky and atmosphere. In real life, this provides an additional reflective illumination outside the actual light source. When the sky is covered by clouds, this effect increases, while on a clear day, the direct light from the sun is dominant.

Additionally, the color of the light changes depending on the time of day, ranging from shadows of grey and blue at dusk and dawn, to shadows of red at sunrise and sunset, up to a light yellow (near white) in direct sunlight.

**Fig. 110.** Example of a landscape created using standard light sources. The diffuse reflection was achieved here in a very simplified simulation by several light sources.

The scene represented in figures 110 and 111 shows a landscape which, in the course of several animation sequences, has to be rendered with many variable parameters.

For one or several single frames, a radiosity procedure would be a possibility to achieve a high degree of realism. However, time and resources are scarce - the rendering network can only be used during the night – during the day the computers are used as normal workstations.

For these reasons, the outside illumination has to be achieved by a minimum of light sources and computing time.

As a rule, one should only use one light source as the main light (sun). The number of fill lights, used in the following example, is a recommendation for daily use. You may want to test the scenic effects by trying out several fill lights, which may also be used to create colorful illumination effects. Try out the various types of shadow, and, as always: look up the contents in the user manual where everything is explained in detail.

### *Main Light, Leading Light or Sunlight*

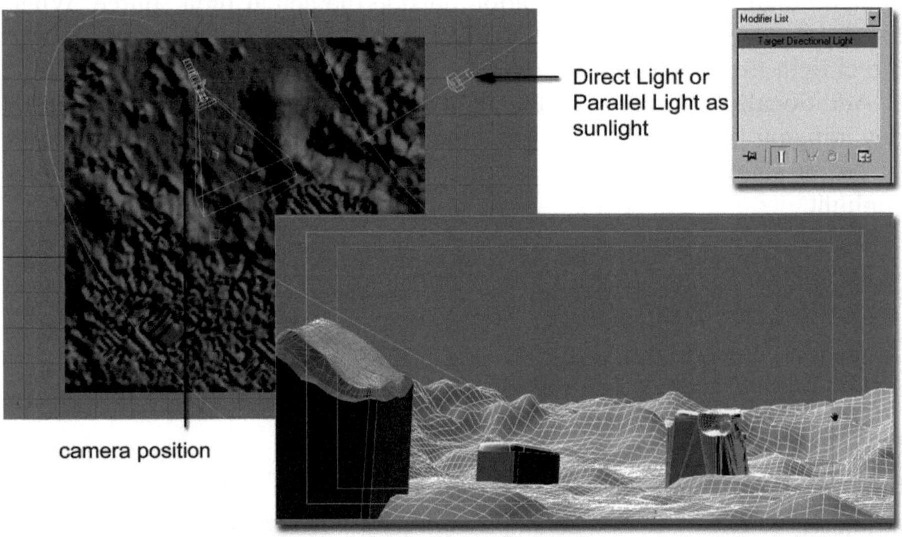


**Fig. 111.** The sun is simulated by a targeted light (direct or parallel light). The applied type of shadow is a raytrace shadow, which causes sharp contour lines.

The sunlight is simulated by a single light source. A parallel light with a raytrace shadow was chosen as the type of shadow. When there are no clouds, sunlight creates a shadow with very sharp edges. This can best be achieved by a ray trace shadow.

### *Backlight*

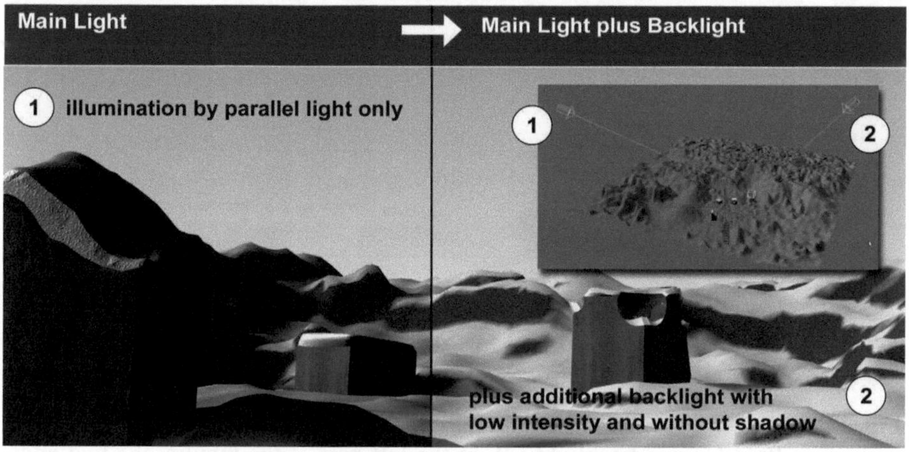


**Fig. 112.** The left half of the picture shows the rendered result with only the main light (1); in the right half, an additional backlight light was activated (2), but without shadow.

The illuminated scene above is still far too dark. The shadow areas are nearly black and the mood is very somber. Therefore a backlight is applied for lightening up the contours. The backlight should be positioned opposite the main light and should not cause shadows. Its light intensity should not be more than 40% of the main light. In the example the, backlight was obtained by copying the main light. The backlight is positioned at the same height as the main light.

### *Fill Lights and Light Atmosphere*

In the following example, the indirect illumination of the sky is simulated rather imprecisely, but this should be sufficient for everyday practical demands. Two additional fill lights are added. Point light or omni light is chosen as the type of light. In this instance, the second fill light is copied from the first. This implies that later changes in intensity or color of one light are valid automatically also for the second light.

Both light sources are supposed to throw shadows. However, the shadow may not be as sharp as the previously mentioned raytrace shadow, but very feeble and fuzzy. Therefore the shadow type "shadow map" is used here.

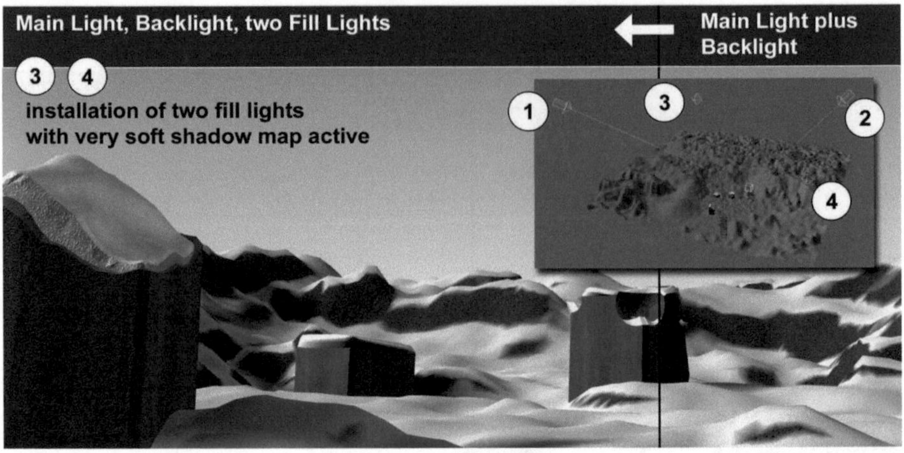


**Fig. 113.** From right to left – the right part of the picture shows the rendered result with main light (1) and backlight (2), in the left part of the picture, the two additional filling lights (3 and 4) have been activated.

The shadow map has a lower shadow intensity than the standard setting, and is set at very soft. In the example this means: reduce intensity of shadow, set intensity on 20% of main light, and set contours in shadow areas on very soft shadows. By doing this the shadows will be noticed but are not dominant.

The two filling lights are positioned slightly lower than the main and backlights. The intensity of the two filling lights should be 10% - 20% of the main light each. These two filling lights are mainly responsible for the illumination of the scene from the sides.

## *Skylight*

The illumination coming directly from above is simulated by another fill light. To a certain extent, this fill light can define the coloring resulting from the diffuse light of the sky. The position of this light source should be a bit higher than that of the first two fill lights.

Here too, the shadow function should be active, the shadow acting similarly to that of the two fill lights described above.

The light source should be positioned a little higher other light sources. The intensity may not be higher than 20-40% of the main light source.

Now, at last, it makes sense to play around with the intensity of the various light sources and to test different values a little.

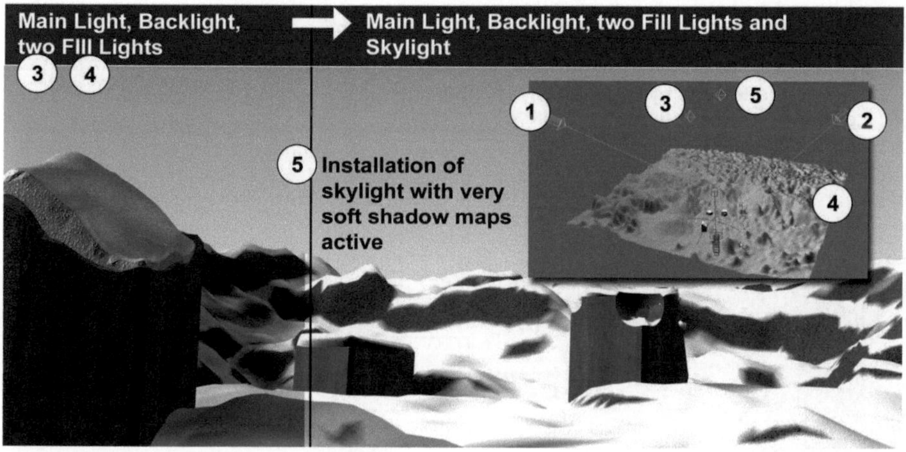


**Fig. 114.** The left side of the picture shows the previous state without, the right side the current state with an active skylight (5)

## *Diffuse Reflection*

At least one more light source is necessary for a small reflection from the ground.

This light, defined as a point light should be situated below the concrete pillars. The downward distance of the light towards Z should correspond to the upward distance of the two filling lights in the atmosphere.

Because this light source is to simulate the effect of reflection, it is provided without a shadow.

It is important to exclude the surface of the terrain from being influenced by this light source.

Furthermore, it is recommended to activate the manual setting of diminishing the light and to restrict the area to the height of the pillars above the ground.

Although the resulting diffuse reflection does not attain the quality of a radiosity calculation, a 3D scene illuminated in this way is quite adequate as a quick illumination and especially for animated sequences.

Incidentally, another important aspect is the data exchange. Should it be necessary to process the model data of a 3D landscape further, radiosity data is normally not saved in export formats. The method of illuminating with standard light sources as presented here, is however interpreted similarly by most 3D programs.

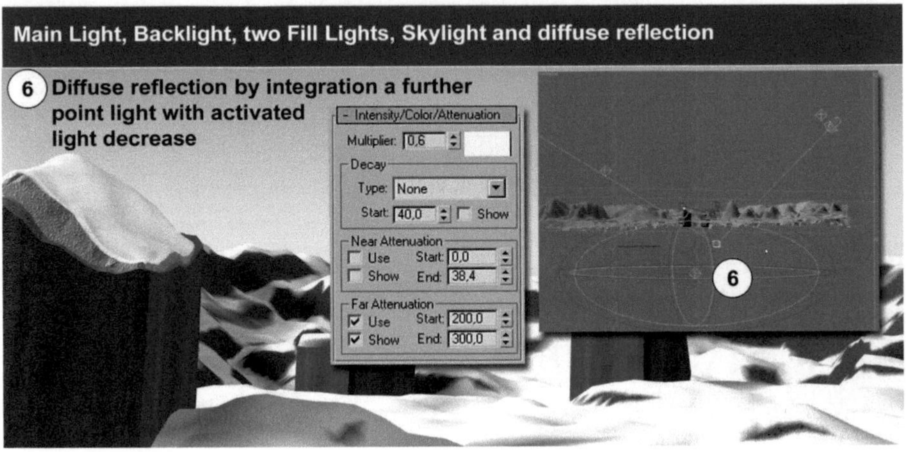


**Fig. 115.** The finished picture with additional diffuse reflection from the ground.

### Tip Re-using illumination

*Should you repeatedly need similar illumination for similar objects, it helps to save the light sources in a special file which can be used as a template if and when necessary.*

## Daylight based on Photometric Light Sources

As opposed to the example mentioned above where five light sources were needed to achieve a more or less adequate result, the same scene will be computed with the help of a GI procedure, namely radiosity.

The geometry remains the same, with the difference that the five standard light sources are replaced by one single photometric light source. This light source is positioned in the same place as the parallel light before. Whereas the standard light sources were defined in RGB values and an intensity of 1,0 (system units) before, the sunlight applied is now provided with physical light values. In this case, the intensity of the sunlight is set at 80.000 lux; the values for the color of the light can, of course, in this case too, only be given in RGB.

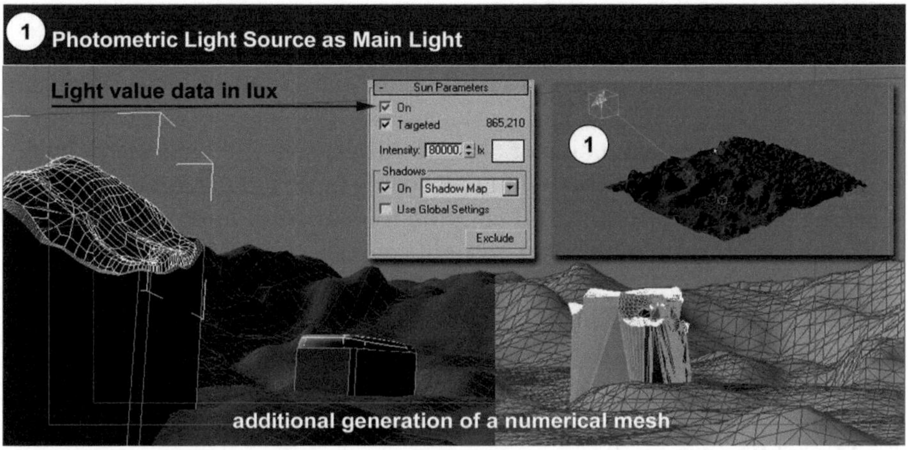


**Fig. 116.** The picture of the 3D scene shows how the calculation mesh has changed after finishing the radiosity calculation.

It is evident in figure 116, what happens after calculating radiosity: an additional numerical mesh is generated. This mesh now serves as a basis for the calculation of light distribution within the expanse.

The advantage is obvious. One single source of light takes over the entire process, which, before, had been achieved using five light sources as an absolute minimum.

**Fig. 117.** The finished picture with a photometric light source

However one should keep in mind that:

- The correct parameters for a GI illumination are no less cost intensive than the settings for standard light sources
- The original file had a size of about 5 Mbyte, the file with the radiosity solution a size of 21 Mbyte (in other words more than factor 4)
- That the calculation of the GI solution took about 8 times as long as the picture calculation with standard light sources

The values presented in this case refer to 3ds max7. The calculations were done with standard light sources and the in-house radiosity procedure. When using the renderer Mental Ray, there may be a reduction in the calculation of the GI solution (provided the respectively optimized shaders for Mental Ray are used).

Basically it can be stated that in nearly all 3D programs, the use of physically correct light distributions requires more time and calculation effort.

## Sun and Moon

When the sky is free of clouds, sunlight is bright, and it comes from one direction. The color of sunlight varies during the course of the day and according to the change of the seasons. The reason for this is the varying depth of the atmosphere which has to be penetrated due to the changing angle of the sunlight and the continuously changing atmospheric conditions. Sunlight is brightest at midday. During dusk or dawn the colors of the sunlight are dominated by red and orange hues.

### Tip Light Color for Sunlight

*Color values for sunlight (RGB) 240, 240, 188, example for shadow color in sunlight RGB 30, 15, 80.*

For simulation at midday, sunlight has a yellow hue and produces a shadow color situated in the complementary color area near violet. Please take note of this when designing a sunlight simulation.

Moonlight acts similarly to sunlight. Only the color and the intensity of the light differ. Reflections and shadow parameters are nearly identical in bright sunlight and in moonlight.

At night only a little light reaches our eyes. The influence of the environment is minimized. Here, the so-called reference light becomes important. The darker a scene has to be, the more important it is to apply the reference light skillfully.

Brightly illuminated windows in a building, for instance, give an idea of the contrast between light and dark. The surfaces of the windows take over the task of the reference light.

## Shadow

The previously presented calculation methods can, up to a certain point, be combined. One possibility, for instance, is to designate ray trace or GI characteristics only to specific objects. So only the main light is provided with the calculation intensive shadow type Raytrace, - while all other light sources are generated by shadow maps.

### Shadow Map

The idea behind a shadow map is, in the first instance, a more speedy picture calculation. A grey-scale picture is rendered. This grey scale picture is only temporary. In this grey scale picture, hidden objects and their shadows are taken care by the position of the light sources. The result will then be "superimposed" onto the picture rendered without shadows.

Incidentally, a shadow map can be separated when rendering, in other words, it can be generated as a file or picture channel of its own. The advantage offered by a shadow map is – depending on the number of light sources – a very fast picture calculation.

However, a shadow map also has a few advantages. They cannot be used for calculating transparency maps, and it cannot represent motion blur either. Working with shadow maps requires a certain degree of experience in order to be able to generate the parameters of the shadow convincingly. The advantage of speed is, however, quickly cancelled by the large number of light sources as well as the high resolution of the shadow map. When the resolution of the shadow is too big, the required memory space grows immensely.

### Raytrace Shadow

On principle, raytrace shadows need more calculation time than shadow maps, but they need comparatively less main memory (RAM) - of course provided that a very high and corresponding resolution of the shadow map was chosen for comparison.

Sunlight, outside shots with an extremely bright light source, or the necessity of using transparency maps, are without a doubt the domain of raytrace shadows.

Keep in mind, however, that shadow generating light sources should only be used in cases where a shadow is required, because shadow calculations demand costly time.

**Table 5.** Types of shadow

| Type of shadow | Advantages | Disadvantages |
|---|---|---|
| Shadow Map | Generates soft shadow contours. When there are no animated objects in a scene, the calculation of shadows is only done once. | Very intensive use of memory. Transparency maps can not be represented. |
| Raytrace Shadow | Supports transparency when using opacity maps. Lower memory capacity than shadow map. When there are no animated objects in the scene, the calculation of shadow is only done once. | Slower than shadow map. Soft shadows are not supported. |

Shadow map as well as raytrace shadow offer the possibility of defining shadow color and intensity.

## Lighting Techniques

Before you start assigning your components for illumination in the 3D scene, it is recommended you think about the composition and structure: plan first, then illuminate!

Here is a short checklist for lighting:

Work with standard lighting. Create your models and compose the scene. Do not use color materials yet.

Put your camera into position, or generate the view perspective.

Assign grey scales to your objects

Render a test picture. When you have the impression that the scene expresses what you want to show, then start to assign your lights.
Place the main light first, and assign shadow to this light.
Render a test picture. When the scene is suitable, start adding one or more fill lights and keep in mind that a fill light has a lower intensity than the main light.
Now start to assign materials and check whether the scene with added color still conveys the same message.
Try to develop a feeling for the position of the light sources in a scene.

By the way, it is certainly not a bad idea to include the illumination into your storyboard.

## Summary

Lighting is a subject that can fill a book on its own (there is one on the market already, by Jeremy Birn, which I strongly recommend to anybody interested in knowing more).

This chapter has dealt with some basics of illumination in 3D programs, presenting terms, procedures, methods and possibilities.

As with everything else, for the real life work situation there is only one option: practice!

Light can be classified according to kinds of light or according to function. Kinds of light are

Point Lights (Omni lights)
Target Lights (Spotlights)
Direct Light (Parallel Light) and
Area Lights

The classification according to its function distinguishes between:

Ambient Light,
Main Light (Key Light or Guiding Light),
Backlight and
Fill Light.

Which illumination procedure is chosen can be decisive for the speed of editing and the quality of the result.

There is a fundamental difference between physically "correct" calculation procedures which take into account aspects like the decrease of light and diffuse reflection, and other calculation procedures which follow more simple models.

The term "illumination procedure" refers mainly to Local Illumination (LI) and Global Illumination (GI).

An excursion into "real life" provided more insights into the sun and the moon as sources of light.

In an example, a near-natural illumination was presented, using standard light sources only, without a physically correct light simulation (e.g. Radiosity).

There are different types of shadows. The most important ones are Shadow Map and Raytrace Shadow.

Lighting is of the utmost importance, and a brief checklist may help you to remember the relevant points.

# Vegetation

*Vegetation can dominate a landscape, as it is the first thing to meet the eye. For the designer of convincing 3D scenes, plants are the first encounter with the challenge of visualizing natural phenomena.*

## Introduction

Do you like going for a walk? If so, you will certainly have noticed that vegetation makes landscape look more complete. One can do without, but with vegetation, the landscape can look so much better.

A walk through the woods is always fascinating due to the incredibly strong presence of trees, shrubs, and grasses. The unbelievable chaos of the bark and its structure hits us at eye level, and, when looking upwards, the view moves into a jumble of branches and twigs. When in leaf, the tree provides light- and shadow plays, and on hot days it invites you to take a rest.

When leaving the forest or when entering a clearing, it is the grasses that attract the attention and are absorbed not only by the eyes.

Fields planted with wheat, groves at the edge of the forest, the sparse vegetation on the border of dry lands, a row of trees along a river, hedges along an alley, or the shrubs separating the lanes of the highway… If you have ever traveled through the eastern parts of France, or through the provinces of Eastern Germany, you will remember the tree-lined roads which seem to go on for ever.

Plants are not just a marginal feature. They provide food for nearly the whole population of the earth, and their presence around us is such, that, in any culture, they represent a permanent element of human life.

The task of visualizing plants is for the landscape designer, similar to animating characters for the designer of computer games. You have to be in the top league of knowledge and experience when dealing with this aspect which can make or break a design. This chapter deals mainly with types of plants and their use in 3D design. We are looking at the representation as well as at the use for different purposes.

Plants are huge resource eaters, if they are to look convincing. Some tricks for visualizing these insatiable monsters will be presented in the following.

**Fig. 118.** After a fire near Gordon's Bay (South Africa)

## Terms and Definitions

The term "vegetation" is derived from the Latin word "vegetare" – to grow, to fill with life. It is the botanical term designating the entire plant community of an area.

A plant community is a group of plants with the same or similar ecological demands. There are different plant communities in different zones around the world. These can be roughly divided into tropical, subtropical, temperate, and arctic vegetation zones, which can then be subdivided further.

The term "landscape", in common language use refers to the physical appearance of a region. This is characterized by natural factors such as its geographical position, climate, vegetation, and by human factors such as settlement, agriculture, and traffic.

### Demands and Standards

When representing vegetation in a virtual landscape, the modeler will be faced with various problems. Vegetation is made up of a chaotic accumulation of elements. These elements may be composed by very complex geometries. Apart from this, the factors time and dynamics play a very important role when looking at the manifestations of vegetation.

**Fig. 119.** Forest landscape in the upper Rhine area in fall

While it is possible to represent a building by a few thousand polygons and simple geometry, plants cannot be created in this way. Not a single part of a plant is flat, square, or cone-shaped. The straight lines and shapes of industrially produced objects will not be found when looking at the appearance of plants or vegetation.

Billions of polygons would be needed to represent a simple tree more or less realistically. Furthermore, with trees and vegetation, a dimension of time is introduced into landscape models: growth, flowering, fruit bearing, adjustment to the position of the sun and to the predominant direction of the wind - all of these are phenomena which are far too much to handle for any statistical model. Apart from this, leaves rustling in the wind add an element of sound to the landscape.“[1]. Fortunately, the human eye is prepared to accept simplified models.

The question is, which areas must be represented, and which areas may be neglected without ending up with a bad result.

[1] Stephen M. Ervin: Agenda für Landschaftsmodellierer - Steine auf dem Weg zum Weltmodell, Garten + Landschaft, 1999/11.

When looking at the edge of a continuous cover forest (as an example for a more complex plant community), one can distinguish between three different levels of vegetation:

The tree level, including smaller and bigger trees
The shrub level, consisting of shrubs of varying size
The groundcover level, made up of diverse grasses, herbs and other ground cover

Except when planted along avenues or in orchards, single plants are seldom to be found in a landscape. This fact leads to an additional problem when designing a virtual model.

"Groups of plants represent a special category of problems. Trees in the far background show different modeling characteristics than those in the foreground. In order to obtain an improved or more easily controlled forest model, the representational authenticity of single trees is often sacrificed. The same problem is encountered in the leaves and in many other vegetational structures. On a small scale of representation these form a mass, on a large scale they are showing individual forms".[2]"

## Why Vegetation?

It is safe to say that, when trying to design a realistic-looking landscape, plants cannot be avoided. The examples of use as well as the fields of application are manifold and varied. While in the field of construction planning, plants are used more as a supporting element; for increasing the authenticity of a scene, the landscape planner will impose much stronger demands when having to represent plants as close to reality as possible.

If a background for a computer game is to be generated, the plants have to look convincing without needing to be botanically correct, except in the case of a game for explaining botany. But in any case, the plants should look acceptable.

If plants are needed for a more complex composite project, or even for a movie, "near-reality" will not be sufficient. Such a case demands photorealism. Photorealism is the little bit of extra time and effort which means that, after the first eight weeks of work on the project during which 95% of the entire project is finished, another eight weeks have to be invested for finishing the remaining 5%,- in order reach the goal of photorealism.

---

2 Stephen M. Ervin: Agenda für Landschaftsmodellierer - Steine auf dem Weg zum Weltmodell, Garten + Landschaft, 1999/11

One has to agree on how much effort is to be invested in the project at hand, and find a suitable model for fulfilling the requirements. What degree of reality or "authenticity" is needed in order to achieve a convincing scene? This question is not easy to answer straightaway, because the decision will have to be made for each case individually.

## Where does the Information come from?

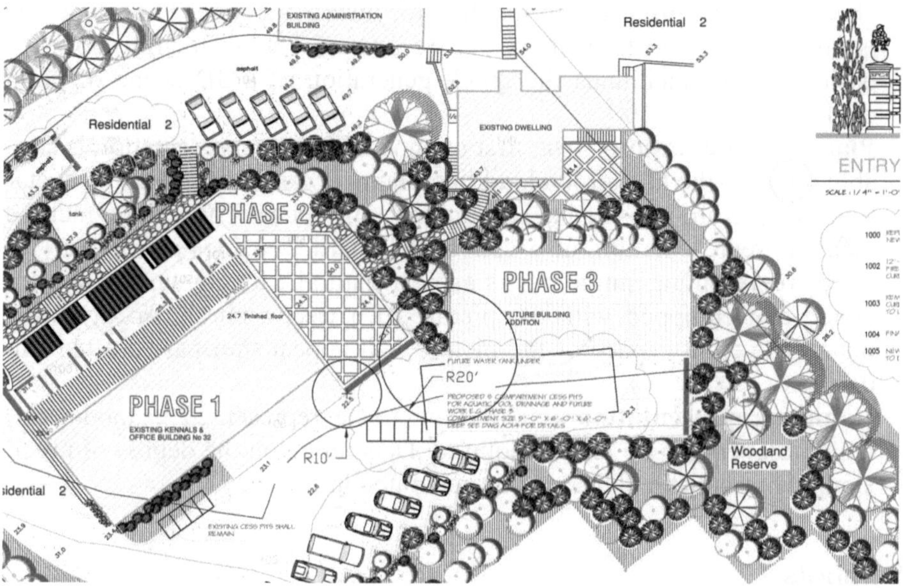


**Fig. 120.** Example of a construction plan including planting plan. [3]

As in nearly every landscape planning project, the visualization of the plants is based on a sketch with the corresponding planting plan.

In most cases the landscape architect will draw up plans in various degrees of detail, often including detailed views and sectional views, giving a sufficiently clear idea about the intended planting projects.

Normally, the position of the plants is indicated by a specific symbol on the map.

---

[3] From AutoDesk Civil 3D Sample Data, SPCA Site Plan

## Types of 3D Representation

There are different ways of visualizing plants and vegetation. These can be classified according to the different design methods. The most popular types of plant representation can be described as follows:

**Symbols** – Instead of trying to visualize real plants, simplified symbols are used. The quickest and easiest way of generating a symbol is by choosing basic, primitive forms, like for instance a cone. The advantage is obvious: this saves memory and resources. The abstraction down to basic symbols facilitates the quick generation of a 3D scene of good quality

**Plane representation** – The first method of plane representation is suitable only for the background, adding plants or zones covered by vegetation as a picture into the background of a 3D scene. The second method of plane representation, also called "billboard", enables a first step towards representing single plants as near natural. Here, a surface is generated and equipped with the picture of a tree, a shrub, or any other plant. The empty space of the picture will appear transparent after rendering.

**Volume representation** – The plant to be represented is modeled – down to every single leaf if required. Depending on the degree of precision, photorealistic plants can be generated in this way.

### Symbols

Using symbols for the representation of plants may be a matter of personal taste, but it certainly makes sense when having to represent the vegetation of a landscape quickly and effectively.

**Fig. 121.** Simplified symbols as plants

In the representation of models, realism is not always needed. A simplified method of representation, as can be done with symbols, in many cases enables the modeler to convey the important contents quickly without the unnecessary "burden" of a costly visualization. In this case, it is best to follow the principle of reduction and less is more.

## Plane Representation

### *Pictures for the Background*

The first method of plane representing is exclusively aimed at generating backgrounds. In most cases a photograph or a painted picture (Matte Painting) is used to compose a suitable background for the scene. This method is excellent for large forest scenarios in the background.

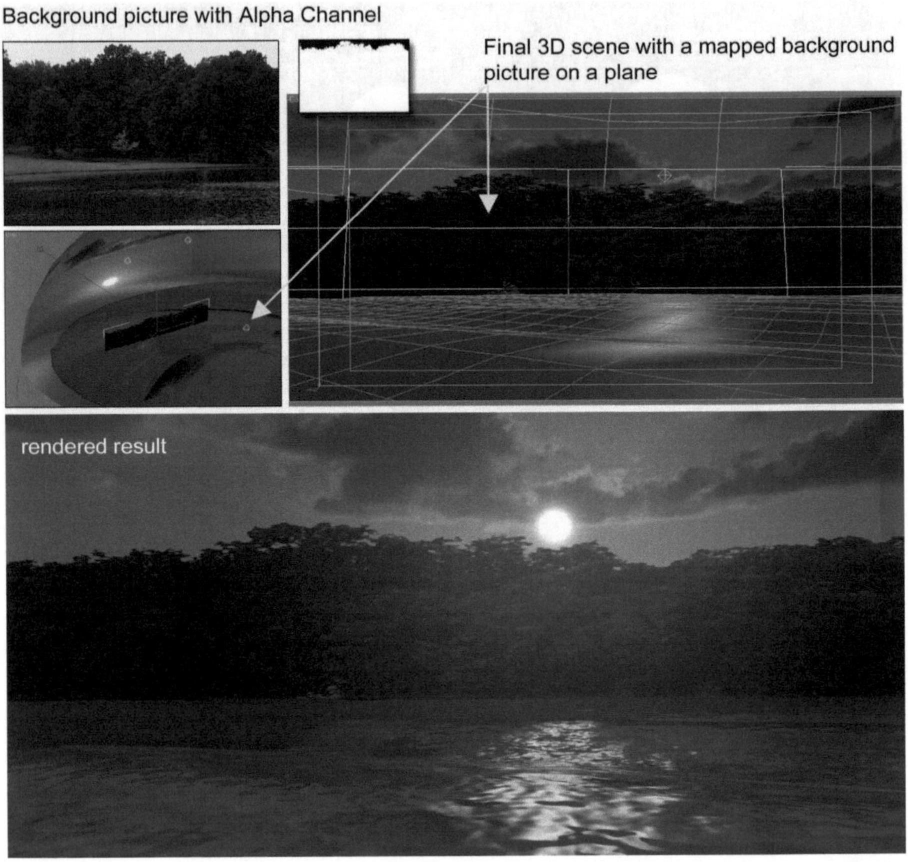


**Fig. 122.** Background picture with alpha channel as texture on one level for a simplified representation of a forest background

The example shows the silhouette remaining as an area in front of the background. The area to which the forest was assigned reacts to the existing light sources like all other objects in the scene.

## Billboard

The billboard method is another simple plane or surface-based method for generating realistic-looking plants. It should be noted: the greater the distance to the represented plants, the more realistic these will appear.

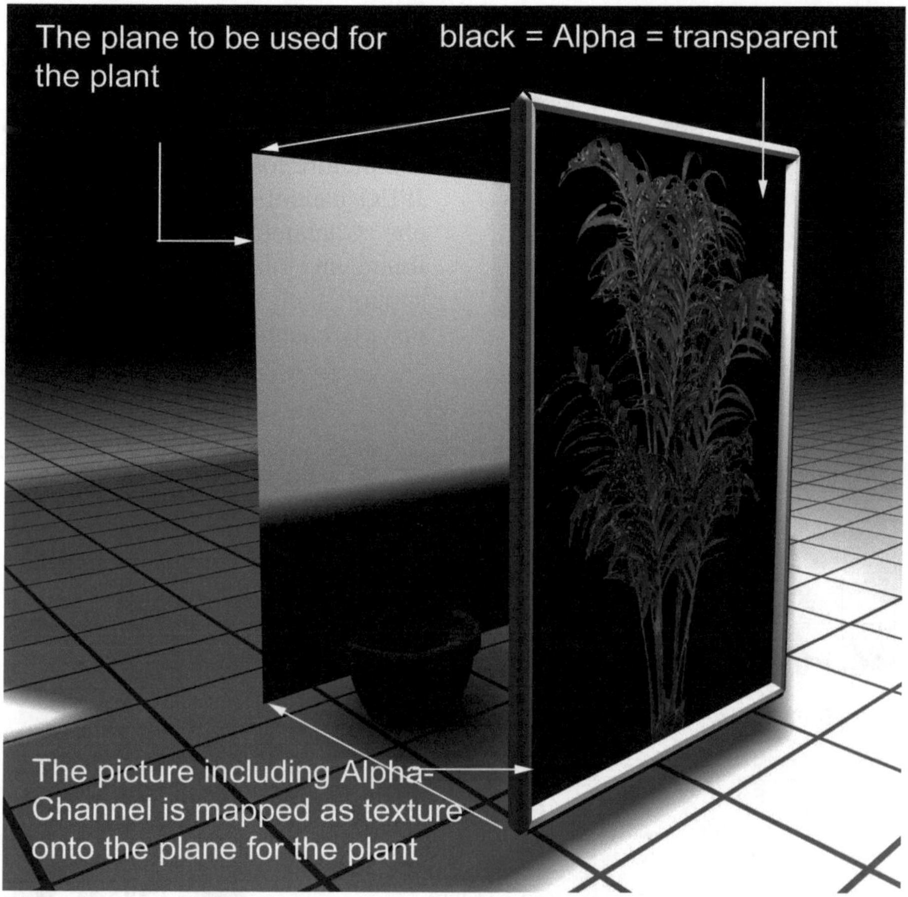


**Fig. 123.** Picture with transparency information as material

The idea can be explained quickly and easily: The plant that is to be represented at your disposal in a regular picture file and provided with an Alpha Channel. The Alpha Channel adds transparency information to the picture. By using an Alpha Channel, certain areas of the picture, depending on the grey scale values, will appear completely transparent or only slightly transparent.

The image is mapped as a texture on a simple surface in the 3D program.

During rendering, the area equipped with transparency information by the Alpha Channel is calculated as transparent. In other words, it can not be seen.

More details concerning picture formats and their characteristics can be found in the chapter "Data Distribution and Post Processing". If a picture format like for instance JPEG cannot support its own Alpha Channel, a method can be found in most 3D programs to support their own transparency or opacity channels.

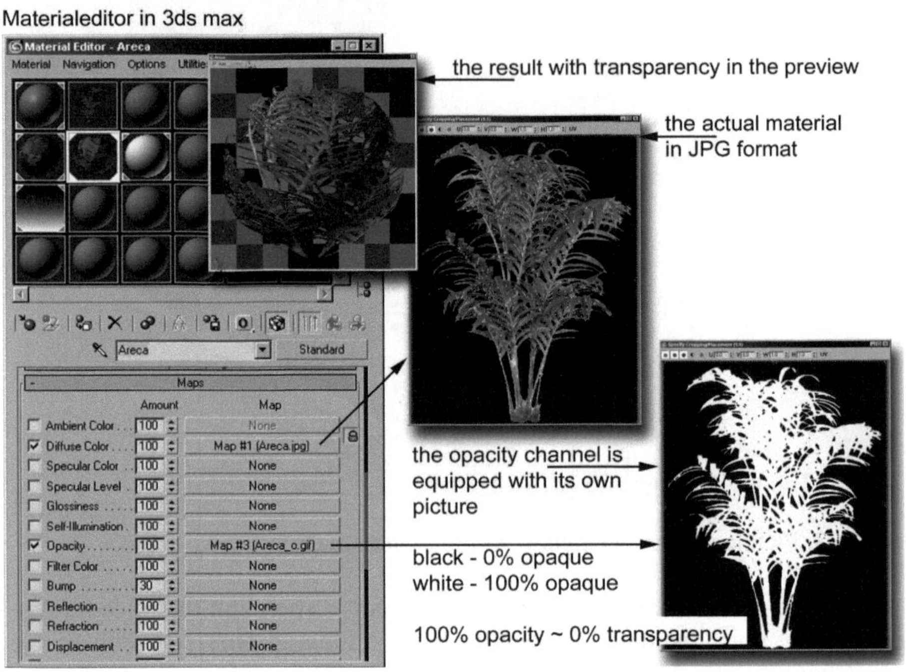


**Fig. 124.** Creating transparency by a so-called Opacity Map

For the texture to be used, a grey scale picture is created which takes over the function of the Alpha Channel. The advantage of these opacity channels lies in the fact that not only the information of the grey scale picture can be used for visibility, but within the 3D environment the intensity of the opacity channel can be modified separately.

**hint Picture Formats for Textures**

*Not every picture format is suitable for use as a texture with transparency information. The formats supporting an Alpha Channel are among others:*

TGA- Targa – Truevision Fomat
TI(F)F – Tagged Image (File) Format
PNG – Portable Network Graphics
PSD - Photoshop – original format

### *Quality of the Masks*

An important criteria for a convincing presentation of the pictures used is the quality of the mask.

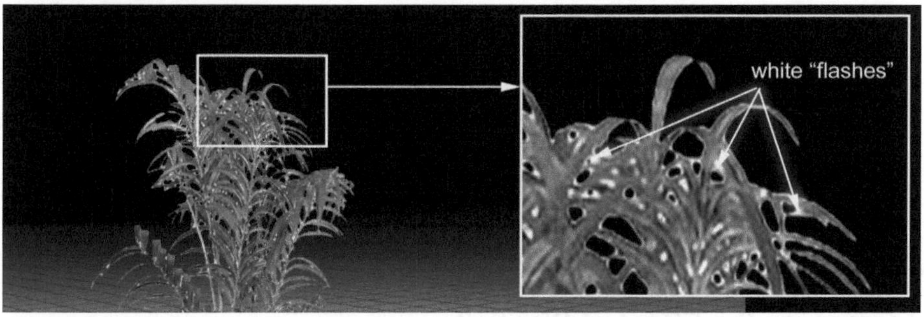


**Fig. 125.** Masking the picture information

Masks are usually generated in picture editing programs like Photoshop. Depending on your level of experience and the quality of the picture material used, the procedure will be more or less straight forwarded. The aim is to avoid the conservation of areas which are supposed to become transparent so that irritating "flashes" do not spoil the image.

### *Billboard and Shadow*

Even though billboards are a quick method for creating the impression of near-realistic plants, there are a few important things to watch out for.

**Fig. 126.** The left picture shows the shadow generated by a shadow map; the right picture shows the same scene with a raytrace shadow

The relatively quick shadow type in 3ds max (Shadow Map or Soft Shadow) is, for instance, not able to represent transparencies. In this case, the shadow type Raytrace must be used, which may on occasion increase calculating time negatively.

### *Increasing the Effect of Plasticity*

A rather simple and quick method to increase the quality of representation in billboard representations is the use of another plane. In this case the existing surface is copied and turned by 90°.

One could be tempted to use even more surfaces. However, this would not result in a substantial improvement; therefore the number of surfaces used when working with Billboard types should not exceed two.

**Fig. 127.** Billboard with a second plane for increased plasticity

## Representation of Volume

In this context we are talking about creating 3D models in a 3D program as realistically as possible. The models can either be designed "manually" by employing polygons, free form surfaces, or patches, or, alternatively, they can be generated procedurally via the respective algorithms.

In the procedural generation, appropriate calculations are used for simulating plant growth. These procedural methods usually generate polygon models.

Deussen describes different ways of generating 3D programs. These are procedural models, regulation of the development of branching, representation via particle systems, and fractal tree models. It would exceed the scope of this book to describe the abovementioned methods in more detail, but to those readers who want to get more familiar with the details of plant growth and the various ways of modeling them, it is recommended to consult the book by Oliver Deussen, which gives an in-depth look into the complexities surrounding plant simulation[4].

The advantage of volume modeling is without doubt the high quality of reproduction of plant details. The problem is, however, that high quality has its price.

The sheer number of polygons of a scene, or even of a single plant, can quickly make relaxed work impossible and turn the composition of a scene and rendering times into a real pain.

[4] "Computergenerierte Pflanzen", Deussen, Springer Verlag, ISBN 3-540-43606-5

**Fig. 128.** Trees generated by polygons. All three trees were designed with the help of scripts in 3ds max [5]

## *Off-the-Peg Solutions*

The experienced modeler will not find it difficult to generate plants within his 3D environment. However, such a high level of expertise can only be reached by a laborious and costly uphill struggle. It makes sense, therefore, to fall back on one of the products that are on the market for generating plants.

These products are based on different approaches and operate differently. Their use depends, like everything else, mainly on personal preferences, and of course on the price.

The approaches range from mere modeling tools like Verdant or Xfrog to prefabricated libraries mostly available in current 3D formats and therefore suitable for direct import into the respective application. Generally these libraries do not offer any options for simulating growth or for animations, but can be used quickly and efficiently.

Most plant modeling tools continue providing direct plug-in solutions for current 3D packages, like 3ds max, Softimage, Maya, etc.

[5] The scripts used are Max Tree Ver. 1.1 Psanjay@rediffmail.com and „Laubbaum“ by M. Wengenroth, www.ilumi.com

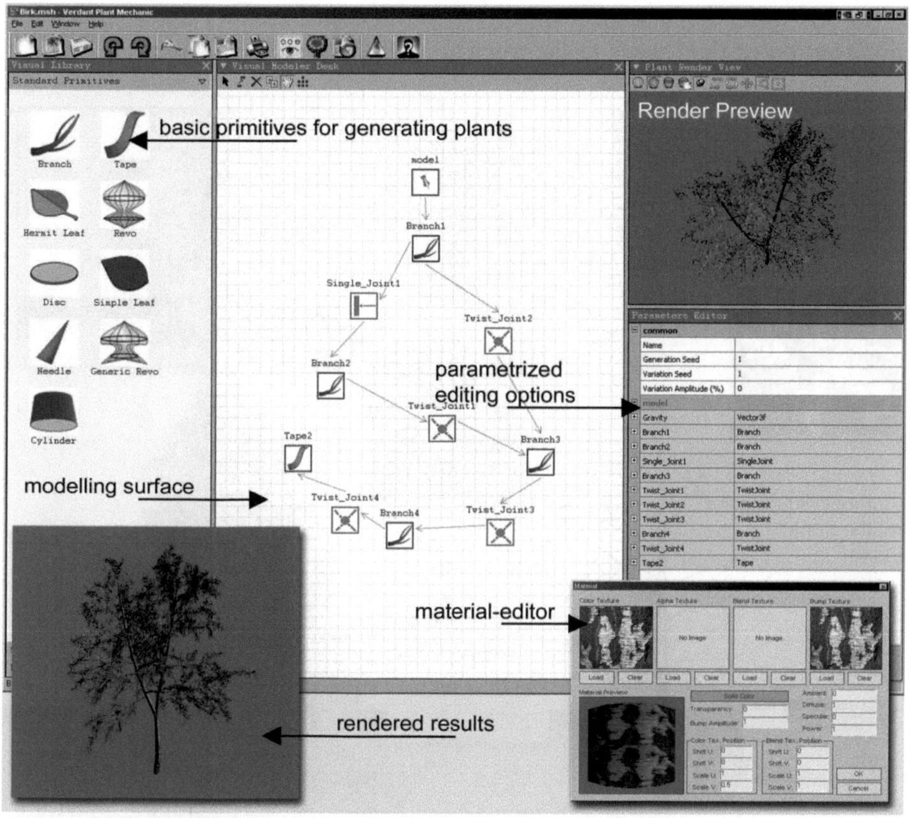


**Fig. 129.** Tree generated via polygons in the plant editor Verdant by Digital Elements.[6]

However, when generating the leaves at the latest, one will encounter the technique of transparent surfaces once again. Because, in order to minimize the polygon output when creating leaves, only a texture with an Alpha Channel or a Transparency Map (as for the billboards) is applied. This way it is possible to use simple geometries, like for instance simple surfaces, for modeling the leaves, whereas the trunk and the branches of the tree are generated in a "real" 3D model.

The figure on the previous page (Fig. 128) shows a tree trunk with branches which have been modeled. The leaves consist of simple polygons with a transparent texture.

The leaves are normally assigned to the tree object by the respective distribution function.

[6] http://www.digi-element.com/

In 3ds max, for instance, this function is called Scatter and enables the distribution of the leaves to the trees according to various control criteria.

**Fig. 130.** Polygonally generated tree with leaves, in 3ds max. The leaves are simple polygons to which a texture with an Alpha Channel has been added.

When a single trunk with branches already has 63442 polygons, the whole tree including leaves boasts with the sum of 121842 polygons. Therefore one should avoid, if possible, this method for modeling entire forested areas.

The advantage of a mixed technique for representing plants is that the shadows will look more or less realistic, even when using soft shadows like shadow map.

Another beautiful example for plant modeling is the use of so-called L-Systems directly within a 3D environment[7].

---

[7] L-Systems were named after the Hungarian biologist Aristid Lindenmayer (1925 – 1989). With this "language", first rate growth processes and selfsimilar fractals can be generated.

## L-Systems

The Blur Studios[8], for instance, have written a plug-in for 3ds max which is easy to handle. This plug-in uses L-Systems for generating any fractal geometries.

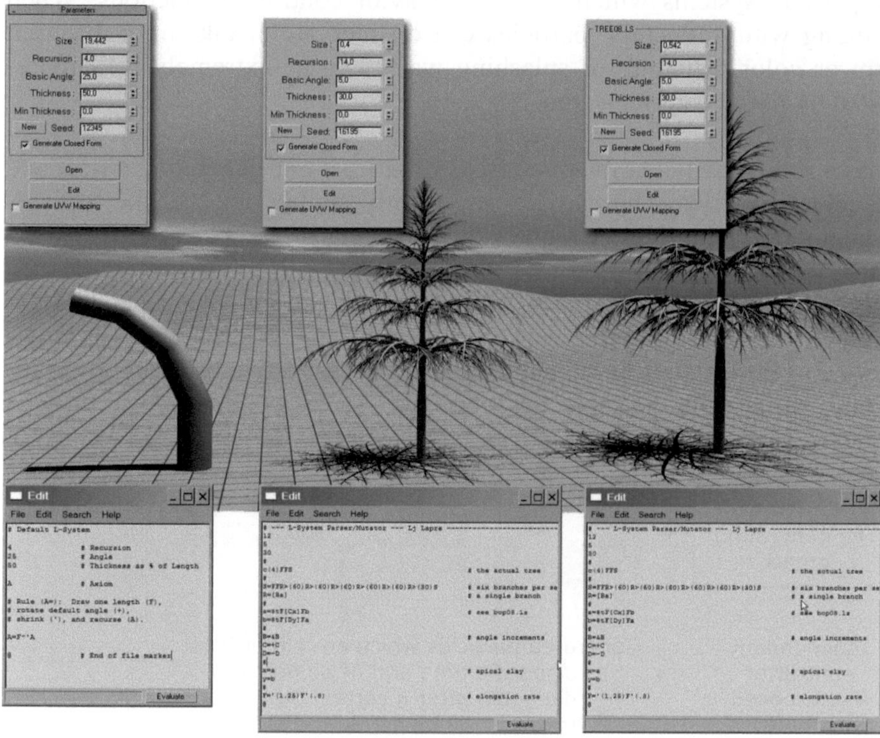


**Fig. 131.** L-Systems via plug-in by Blur, integrated into 3ds max. The change in growth behavior is achieved by entering the parameters into a text window

### Tip Search the web

*It is recommended to look for L-Systems or Lindenmayer Systems in one of the current internet search engines.*

[8] http://www.blur.com

## Particle Systems

Most 3D programs provide particle systems. Usually, particle systems are used for representing rain, snow, or volumetric atmospheric effects. There are particle systems which include behavior controlled functions. When colliding with deflectors[9] particles can disintegrate or take on a different form or color. Running or splashing water can be extremely well represented by particle systems, too.

As a rule, a particle system consists of at least one generating object and the actual particles. The particles are mostly simple objects, like surfaces or geometric primitives, which often look "plausible" because of motion blur.

The generating object, the so-called emitter can be an independent aid solely for generating particles, or any kind of geometry. The use is described in more detail in the handbook of the respective program, or in tutorials dealing with the subject.

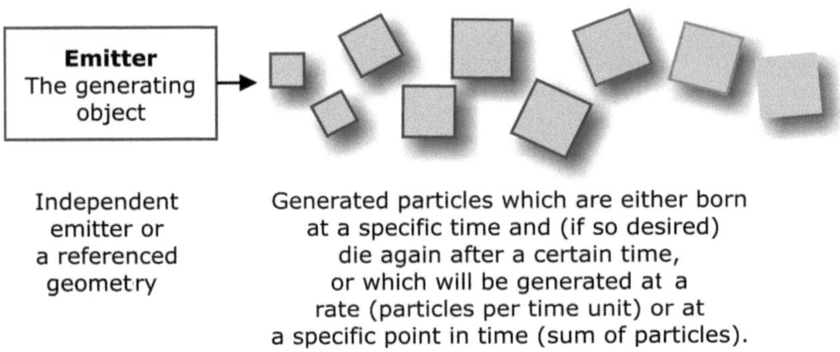


**Fig. 132.** Particle generation of a simple particle system

The advantage of a particle system is that you can control the time at which the particles appear. The generated particles can be "born" at a constant rate within a certain amount of time, or a specific sum of particles can be generated within a specific time.

All particles can appear at the same time as well. Particles can "die" or "live" forever.

Therefore, by using one geometry as an emitter, and another geometry as a leaf, simple tree structures can be generated relatively quickly.

[9] Deflectors are objects specially designed for reflecting particles. If an object is not defined as a deflector, it will be penetrated by particles, which can lead to strange effects.

The following figure gives a simple example for the use of particle systems. A first particle system uses the branches of a roughly modeled tree as an emitter. A polygon is used as a particle which serves to generate further branches. Beforehand, the specific number of particles to be generated is defined.

The branches generated in this way are again used as emitters for the leaves, and, there you go,- the tree is finished.

Naturally, the example does not look exactly like to a real plant, but it demonstrates how quickly simple tree structures can be generated with the help of a particle system.

However, the more detailed the respective geometries are, the more calculation time you will need.

Depending on the field of use or if you are uncertain, it is recommended to rather limit the number of leaf structures and maybe also to refrain from adding additional branches.

Another advantage when using particle systems should be mentioned: they can be extremely easily influenced by outside forces, like for instance simulated gravity or wind.

Basically, a particle system can be used like the Scatter function described above. However, there are far more variation possibilities and editing options.

Furthermore, particles need fewer resources than, for instance, copied geometries, for generating the same effect.

As a rule, particles are represented at rendering only, during editing simplified dummy representations are used instead, in order to avoid making too high demands on the graphics memory. In this way, by using particles, even very complex scenes with a high resolution of detail can be realized.

**Fig. 133.** A possible way of adding leaves to a tree

## Grassy Surfaces

The strong point of most programs for designing landscapes is the generation of general overviews. Whether the programs are named Bryce, Vue d'Esprit or World Construction Set, it doesn't make any difference; in a close-up the scene represented will in most cases lose its credibility.

The grass may be green, but that's not all there is to it. For objects in the distance, grassy areas can be created without problems with a suitable texture. In this case the color green is actually sufficient to achieve the desired effect of a green area covered with grass. However, for a close-up this method of representing grass is not sufficient. As described above, more detailed models are needed for constructing convincing scenes.

The main problem in a close-up is the reconstruction of the chaotic appearance of Mother Nature. Grassy areas especially will soon appear unconvincing when every blade of grass has the same length, and they all look like soldiers standing neatly in a row.

An example of a simple grassy area could look like this:

**Fig. 134.** The "raw" scene, still without plants or vegetation

When finished, the scene has to give the impression of a complete grass cover.

## Constraints

In the example, the scene consists of a construction resembling a dolmen. In the foreground there is a tree, which does not look very much alive. The scene is to be equipped with grass. The camera is to take a still. A camera movement through the scene is not planned. As an option the grass should be shown to be growing in the course of time, and the influence of the wind on the grass blades should be shown as well.

## Texture

As the camera position will not change, it is not necessary to provide the whole scene with a grass cover. Therefore the first step would be to create a suitable texture for the grass cover in the background. A photograph of a piece of ground covered by grass would be quite sufficient for the first impression.

**Fig. 135.** The basic ground level was provided with a texture

The grass looks quite pleasing, but only as long as the camera is far enough away. In spite of a rough bump mapping, the texture is not adequate for a close-up.

## Modeling a Blade of Grass

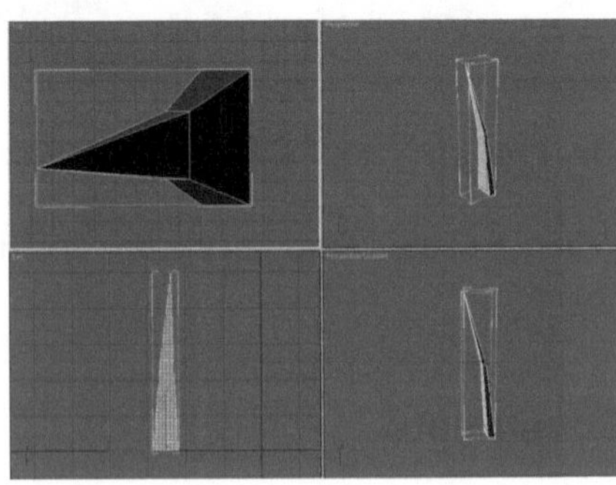

On the basis of a pyramid, and with a bit of manipulation an object similar to a blade of grass is created. It could also have been an object based on a polygon or any other geometric form. What is important in the present example is the impression of a grass blade. Furthermore, it is important to keep the number of polygons as small as possible.

## Growth Areas

Whether the grass blade is copied manually after its construction, or distributed across the surface via a distribution function, or whether a particle system is used, the fact is that the camera is positioned in the foreground, and it does not make much sense to distribute the grass on the entire surface.

The use of a particle system for creating the lawn makes sense due to the additional demands caused by the growth of the grass blades and the influence of the wind on the surface.

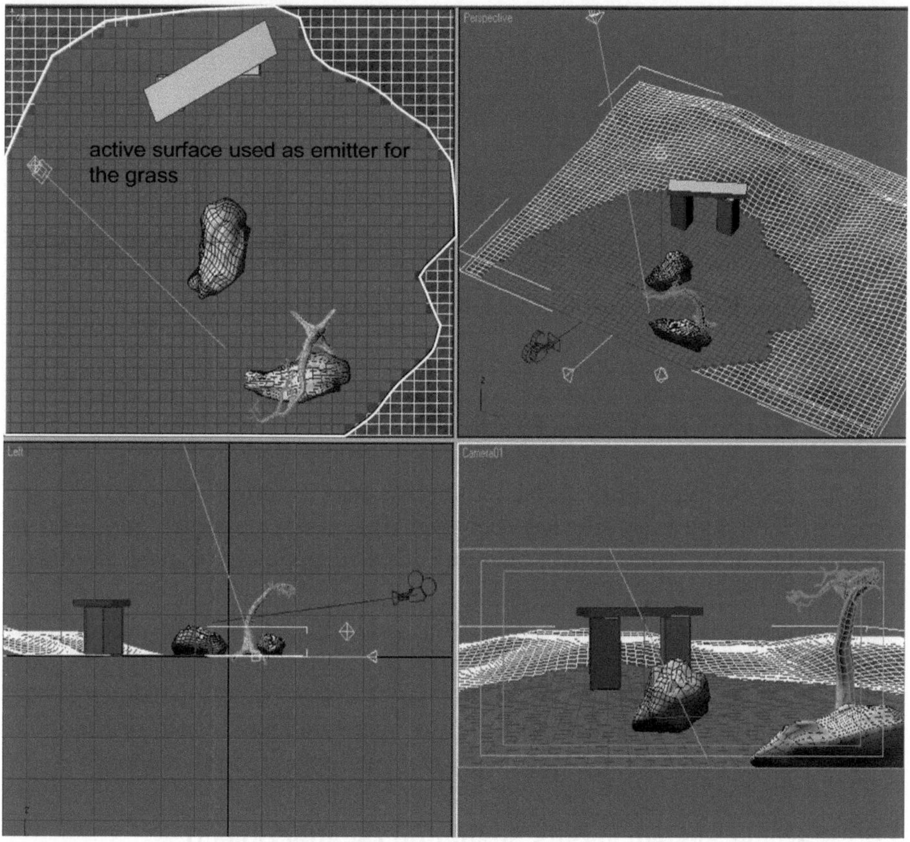


**Fig. 136.** Selection of polygons to be covered by grass

When selecting the polygons that are to be provided with grass, it is a good idea to leave out the rocks in order to avoid an unwanted penetration later on. If this is not done, the grass will, regardless of existing objects, penetrate everything that is in its way.

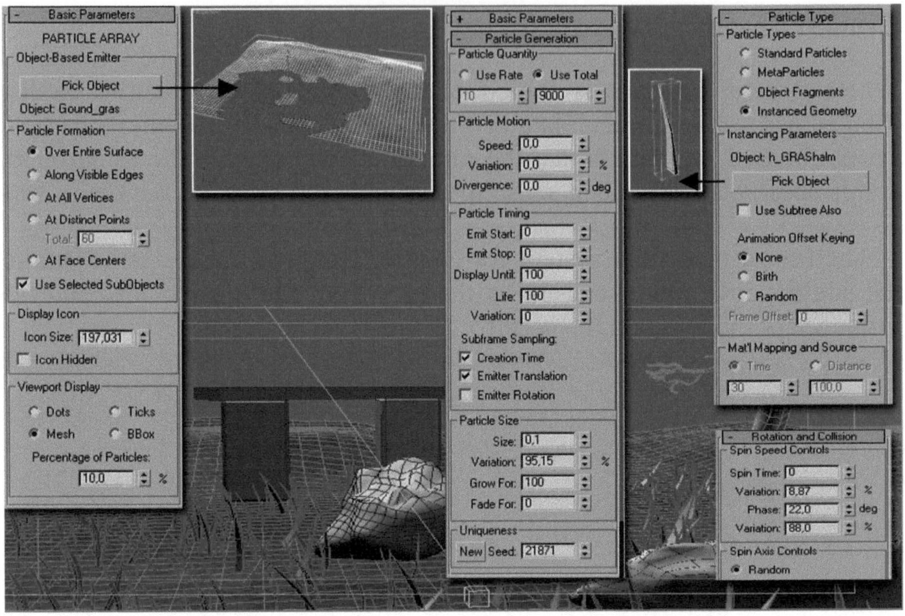


**Fig. 137.** Restricting the selection of polygons to avoid penetration

## Distribution of the Grass Blade

A particle system is constructed. In the 3ds max example, the particle system *PARRAY* has been chosen, because this system enables any kind of geometry to act as an emitter; also any kind of geometry can be used as a particle.

The surface is selected as an emitter, and the option *USE SELECTED OBJECTS* is activated. This ensures that the particles will create the grass only on the selected areas. For the actual creation of the particles a sum (9000) is activated- they will all be "born" simultaneously at frame 0.

As a particle object the blade of grass previously created is chosen.

### hint Particle System versus Scattering (Scatter object)

*3ds max provides a system, which can distribute any object on the surface of another object. For statistical purposes and for polygon numbers that are not too large, this is a suitable and quick solution. In the case of large distributions and if an animation may be needed at a later stage, particle systems are the best choice.*

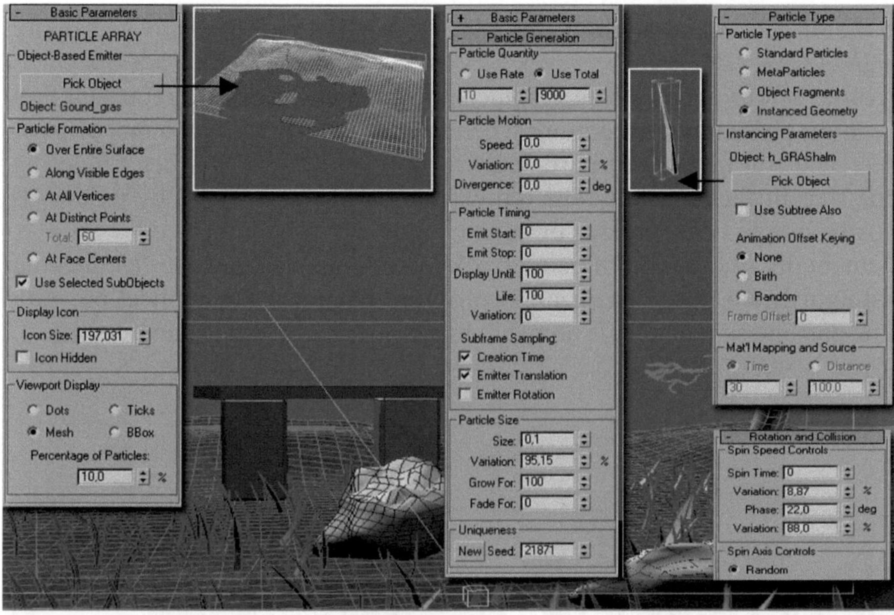


**Fig. 138.** Using the particle system *PARRAY* for generating the grass distribution

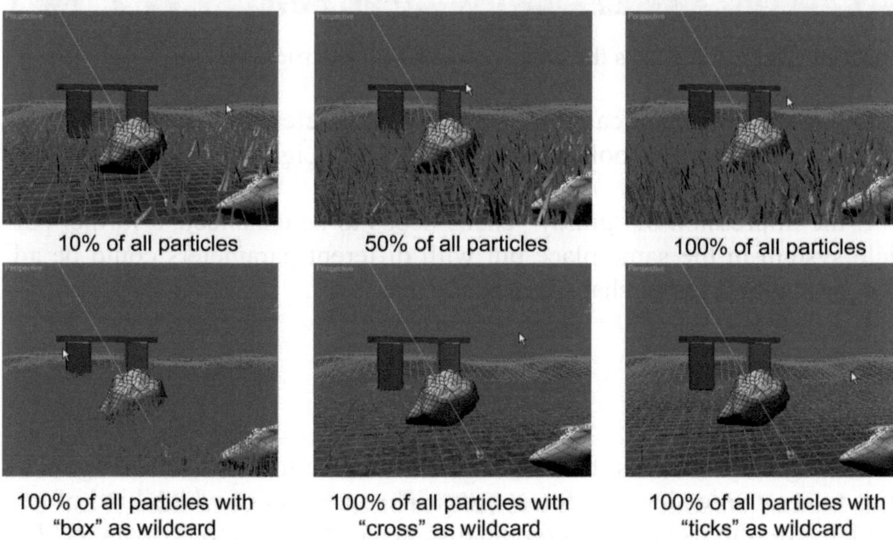


**Fig. 139.** Different representation of particles as seen on the monitor

Because an even surface would appear unrealistic in a grassy area, there should be a certain degree of chaos. A variation of the size (95,15%) will give you an uneven look. A further parameter will take care of a variation

in the direction and angle of the blades: rotation and the activated option *SPIN AXIS CONTROLS*.
In order not to overload the graphics map, it makes sense to represent only a certain percentage for editing when dealing with larger amounts of particles. The full number of particles will only be generated when rendering. One step further would be the representation of the particles by wildcards, mostly simple ticks, crosses or boxes.
Incidentally, the option of using particle systems is one of the features that distinguishes a pure visualization program from a CAD application. As a rule, particle systems have no place in a construction environment.

**Fig. 140.** The result shows the grass generated by a particle system

Although the specifications in the example refer to 3ds max, they are similar in comparable tools like Softimage XSI, Lightwave, Cinema 4D, or Maya.

If the impression of "grassy" chaos needs to be enhanced, a further particle system in the same place but with different parameters could be added, or maybe another shape of particle geometry.

## Forested Areas

As with grassy surfaces, here too, it makes sense to use instanced geometries to create nice looking forest areas. The question of which kind of tree model is best to use depends on the distance of the trees to the camera. The nearer the trees are to the camera, the more important it is that the models used have a high level of detail.

It should be obvious that the detailed generation of a forest sequence, as was produced by Dreamworks for the movie Shrek[10], exceeds the boundaries of a simple landscape visualization by far. Besides, the question here is not the representation of growth models, such as generated for instance by the application Grass[11] in the field of GIS. The procedure for a simple forested area can very well be realized quickly and by most 3D programs.

As with grass, an area is defined as an emitter and subsequently provided with a specified number of particles. As a referenced geometry , instead of the grass geometry, in this case a surface with a tree map and the corresponding opacity map is used.

It is obviously very important to remember to represent the surfaces to be used in as many different ways as possible. In most cases this variance can be achieved by the respective distributions in scaling or rotation.

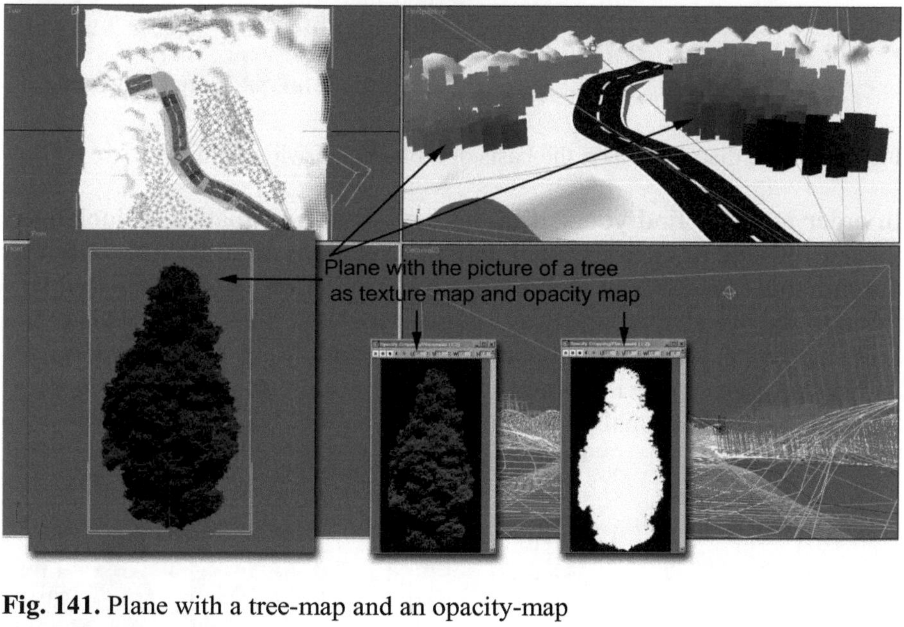


**Fig. 141.** Plane with a tree-map and an opacity-map

Of course, there are ready-made tools for this procedure, such as the program "Forest" issued by Itoo Soft.[12]. The producers not only considered all possible distributions of forest stands, but also provide their own shadow type named Xshadow, which combines the advantages of the Shadow-Map with the elegance of a Raytracer.

---

[10] Shrek Part 1: www.shrek.com
[11] http://grass.itc.it/
[12] http://www.itoosoft.com

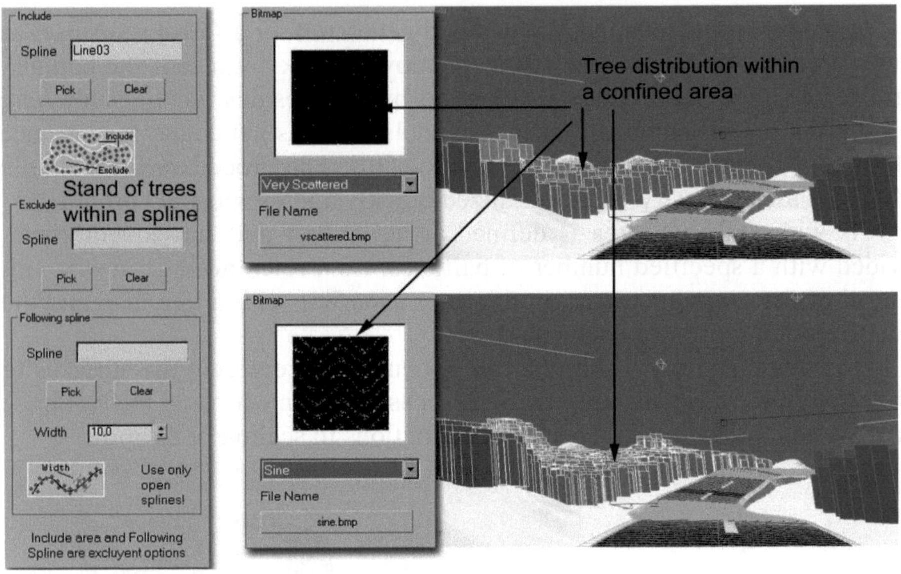


**Fig. 142.** Tree distribution on the basis of a black-and-white bitmap

In order never to find yourself in the embarrassing situation of looking at the narrow side of the area fitted with the tree-map while flying along with the moving camera, it is well worth the extra effort to keep the tree surfaces of the wooded areas always aligned towards the camera. This effect is also very popular for interactive visualizations, a domain where resources are even more important than in the purely statically 3D visualizations for representational purposes.

**Fig. 143.** The road scene with „forest“

**Fig. 144.** Planes representing forest areas that were not aligned to the camera“ by mistake” leave a “flat” impression

For the representation of broad tree-lined avenues, it is preferable to use a “path”, in other words a spline. The trees can be generated along the spline.

## Seasons

The change of seasons is most obvious in the changing vegetation. In winter, everything seems to be empty, not a single leaf is hanging on the tree, the view extends far into the distance, the sound waves are carried much further away than in summer, for lack of the acoustic brakes of the leaves. What is more, under a cover of snow the entire landscape seems to be distorted as well. In spring, there is extremely rapid change. Within a short space of time, colors and forms change. In summer, everything appears in full strength, only to discolor again in fall, to change and to perish.

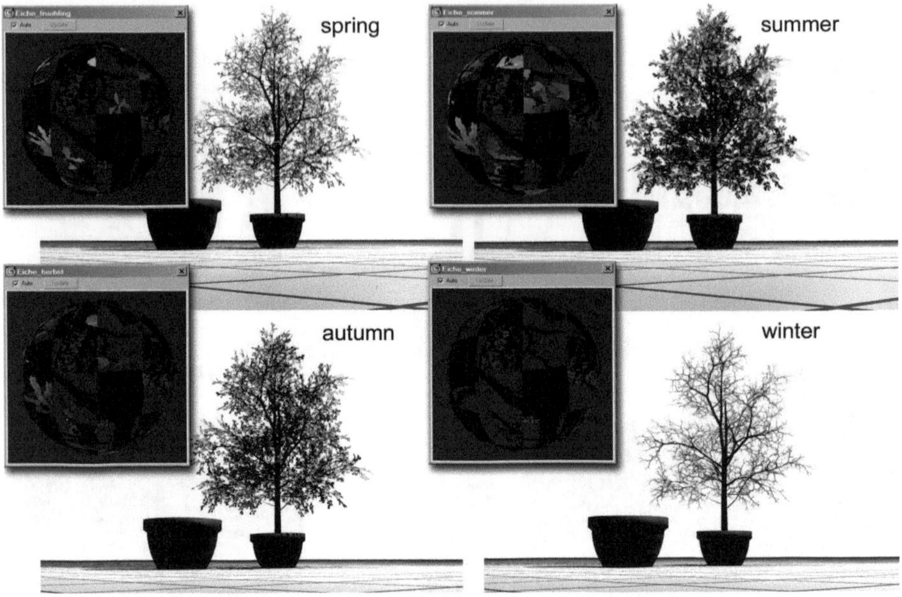


**Fig. 145.** Change of the seasons by different materials

One method to adapt the vegetation as quickly as possible to the season of the year is the generation of seasonal materials. When a material library has already been generated for the summer, it makes sense to use the same terms for further material libraries for each remaining season. Thus, by assigning the respective materials, the mood of the season is quickly achieved.

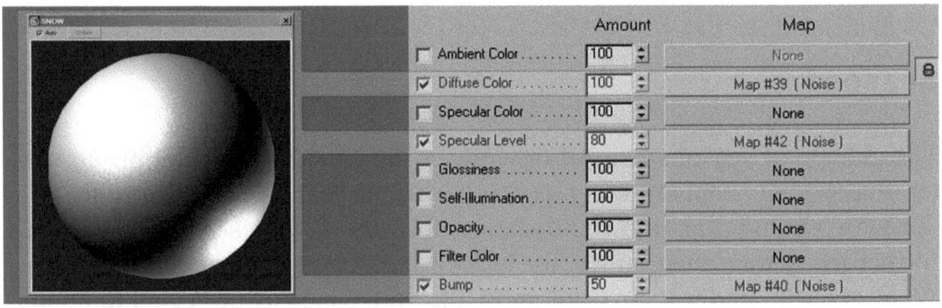


**Fig. 146**. Material for snow: a noise map is assigned to *DIFFUSE, SPECULAR LEVEL* and *BUMP*

Occasionally, only certain details in the picture have to be fitted with snow. The example in Fig. 147 demonstrates a possible method for assigning a snow cover, either partially or completely, to a specific geometry.

The example shows a dead tree, although, this could also be a section of a wall, a bench in the park, or any other object in the scene near to the camera.

Representing winter landscapes it is often a problem how to achieve a convincing visualization. If an entire landscape has to be covered in snow, the effort quickly becomes incalculable. Therefore it is better to concentrate on one detail in the close vicinity of the camera.

First the polygons have to be assigned to the objects that are to be covered with snow later. These are copied into a separate object to which a special, very explicit name is given like, for example, "snow_tree_XY", the areas are extruded, and some noise is added, then, additionally, the mesh is smoothed, and all of this is fitted with a snow material, and bingo, there is snow covering the tree.

This method of generating a snow cover is certainly only one of many possibilities. It has the advantage that snow can be generated quickly and uncomplicatedly on nearly any kind of geometry.

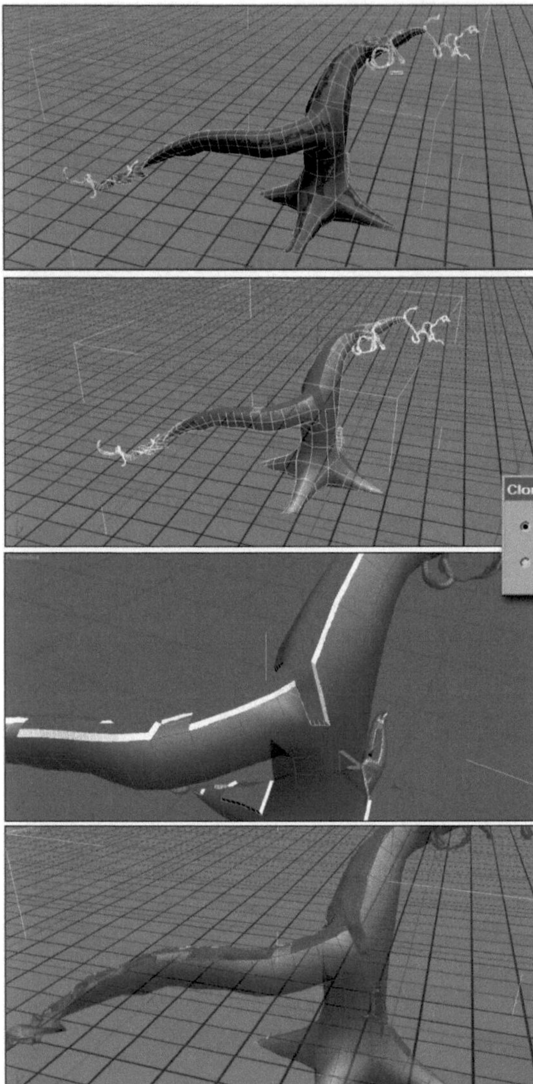

The tree is to be covered partly with snow!

1. Selection of upper polygons

2. Extraction of selected polygons and saving as a new object - here labelled "SNOW"

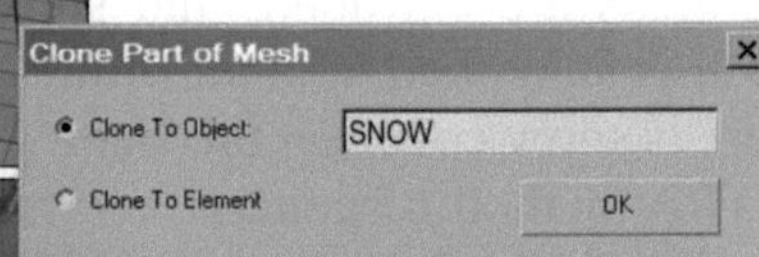


3. Tessalation of mesh elements

4. Loft extrusion of the snow object

5. Adding noise to the snow to generate a "restless" look

6. Smoothing the mesh using "Meshsmooth"

7. Assigning a suitable snow material

8. Rendering ...

**Fig. 147.** Limited snow cover via modeling

# Animation of Plants

## Outside Influences

Often, especially in plants, outside influences play an important role. One only has to think of the wind moving the treetops and causing grasses to bend. A vehicle passing a group of shrubs generates suction, etc. We are seldom conscious of these movements, but when they are not present, the environment appears unreal. The main cause for all of these little movements is the wind. In computer animation this effect is called secondary animation. Secondary animation is also the continuous waving of clothing or the repeated bouncing of an automobile that has just passed a pothole.

But let's not get carried away. We will limit the visualization of landscapes to those aspects which the viewer is still conscious of. The really big efforts for secondary animations should be reserved to the big movie studios with sufficient manpower.

However, even with a small effort the landscape planner can bring quite a few things into motion.

**Fig. 148.** Linking the option "bend" to the referenced geography. The bend function is controlled by the slide control which can be animated

When, as in the example above, the grass is generated via a particle system, it is fairly easy to get additional "outside forces", in this case the wind, to have an effect on the particles. This effect, used sparingly in the vicinity of the camera, will work wonders in convincing the viewer during a walkthrough of a virtual landscape. It is just a pity that the viewer actually only registers this effort when it hasn't been done.

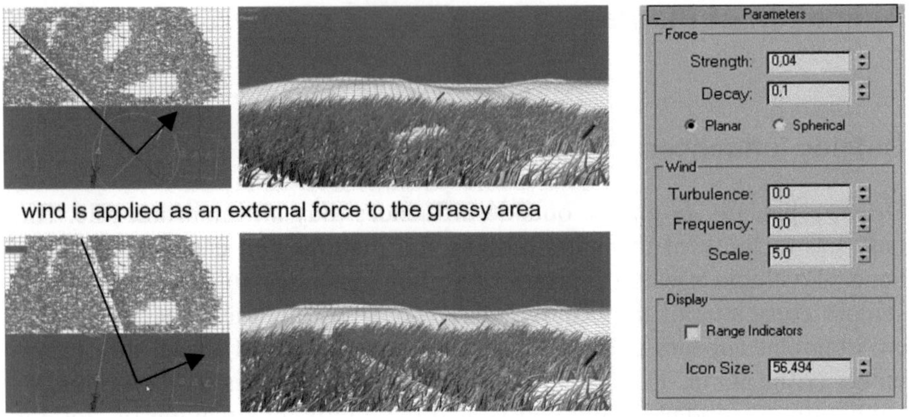


**Fig. 149.** Wind is applied as an external force to the particle system grass.

When dealing with trees the procedure is similar. If the plants in a 3D scene that are to be fitted with a secondary animation, have not been generated via a particle system, there are other ways to animate via a slight distortion (e.g. bending) varied in time.

## Growth

It would by far exceed the scope of this book if we were to occupy ourselves exhaustively with the growth of plants. In this field one quickly comes to a halt against insurmountable features of current 3D tools.

Although 3ds max, XSI, Cinema and co. can offer an incredible variety of tools for animation, the near-natural simulation of plant growth proves to be just too much, even for parameter enthusiasts.

To those who really enjoy this, it is recommended to study the respective programming language (mostly script-based), and to delve into L-Systems[13].

If you do not feel like programming you should take a look at the plant modelers on the market. These offer, on a sound base, a comfortable method of generating plants. Most current 3D programs are compatible with the corresponding exchange formats or plug-ins.

[13] L-Systems see types of 3D representation/volume representation

**Fig. 150.** The particle system grass is fitted with a free growth constant

Although, with the help of particle systems, in this case a convincing impression can be attained relatively quickly, too, the handling of parameters requires a lot of practice and instinct.

**Fig. 151.** Plant growth illustrated by a flower

## The "Right" Mix

Plant modeling is a very time consuming activity. If there is not enough time to do the modeling with custom made tools, it is recommended to fall back on existing tools. It makes no difference whether these are modelers that generate the growth of plants according to various criteria, or simply ready-made libraries for generating billboards or 3D models. In landscape visualization, it is important to get results as quickly and as efficiently and as cost-effectively as possible.

There is no "right" or "wrong". It is rather a mixture of experience and the various methods, which, in the end, will lead to a good result within a reasonable time.

In most cases, the decision is in the hands of the modeler, who has to decide how best to get to a result.

It certainly makes sense to consider the options very carefully and to choose the most suitable tools.

Here are some points to consider which may be of help when choosing the method and the right tools:

**Still frame or animation** – What kind of visualization is needed? A still or rather an animation with a moving camera? For a still, the effort required is considerably less, because only the plant models within direct view of the camera need to be convincing.

**Distance to the camera** – Which plants are in the vicinity of the camera and which plants are in the background?

**Plant species** – Which species of plants have to be generated, and with how much effort? Naturally this consideration includes the question of whether billboards will be sufficient, or whether it is possible to generate or import 3D models.

**Additional programs** – Which ready-made plant modelers or plant libraries are at your disposal, or will have to be bought for this project?

**Rendering time** – What is your computing capacity for generating and rendering complex scenes within the time required?

## Summary

This chapter has dealt with the subject of plants in 3D visualizations from the point of view of the practitioner. First of all, the basic questions concerning the use of plants, and to which purpose they are used, were looked at in some detail.

The various types of visualization can roughly be divided into categories according to their method of generating plants:

Representation by symbols
Representation by planes
Representation by volume

It makes sense to use surface representations mainly for backgrounds. It is important here to use transparency effects (opacity), which can be generated by an Alpha Channel, or by an independent picture file for transparency.

Alpha Channels and transparency information are usually available as grey scale pictures. Depending on the grey value, the picture information is represented in a scale ranging from transparent (black) to opaque (white).

The method used for representing single plants or groups of plants is called billboard.

When using billboards one has to pay attention to the various types of shadow including their respective advantages and disadvantages in the representation of plants. Not all computing methods or shadow types are able to deal with transparency effects.

In real-life examples, different methods for generating grass were presented, with the main emphasis on the use of particle systems.

The same technology of using particle systems is recommended for generating forested areas.

Various methods for the animation of plants on the basis of particle systems were demonstrated, in a short excursion. The main topic was the generation of secondary animations and simple growth animations.

Above all, one point should have become clear: plants mean a lot of work. But it is definitely worthwhile to take a closer look at these aspects because a successful visualization of landscapes is largely dependent on the method of representing plants.

For more background knowledge and further details please consult the literature list. By now, quite a few tutorials can be found on the internet. Since these are mostly limited to the use of a particular software package, it makes sense to look up the respective discussion forums. Generally, most search engines provide good results.

# Atmosphere

*The term atmosphere is derived from the Greek words ατμός, atmós = air, pressure, vapour, and σφαίρα, sfära = sphere. The atmosphere is the gaseous shell enveloping the earth. It consists of a mixture of different gases which are attracted by the gravitational field of our planet.*

## Atmosphere?

First of all, the term atmosphere is used for the complex mixture of gases which, attached to our planet by the force of gravity, makes life on earth possible. There are four layers:

**Troposphere:** the weather layer where the temperature decreases with increasing height, followed by the
**Stratosphere:** in this layer the temperature increases with height up to a value of 0° C, only to decrease again in the
**Mesosphere:** down to -100° C, but then rising again on the level where polar lights are on the rampage.
**Thermosphere:** the last layer separating us from the immense vastness of outer space.

But the term atmosphere is not only defined by these four layers - there is a lot more to it than that. Invariably, it evokes emotions, feelings, and sensations. Nearly always, there is some connection between this term and some non-factual observations.

With respect to the representation of landscapes, atmosphere is synonymous with general mood, frame of mind, or simply the emotional impression which a landscape can evoke in the viewer.

The impact of the message, of a visualization, indeed of any picture, depends to a large degree upon the "correct" rendition of the atmospheric effects. Fog and mist in a nature scene, haziness in the distance, cloud formations, smoke - all these are important elements for generating a convincing product.

In this chapter we focus mainly on the effects which are indispensable for a successful presentation in real life, and therefore also in computer visualization.

Some of these effects take place within a 3D scene, some are defined by materials, and others are generated by so-called Video Post-Effects. These are procedures that are applied on the finished result after rendering, in the form of a filter.

In every-day language, we talk about

Color perspective, and the loss of color towards the horizon
Mist and fog
The sky depending on the time of day and humidity
Clouds
Rain
Snow

Have you ever noticed that

In the morning the color of the air is brighter than at midday
In the evening the colors appear weakest
In winter the air seems to be clearer than in summer
In fall the light is at its "softest"

All of these are very good reasons for many landscape painters and photographers to produce their works primarily in the season of falling leaves.

## Color Perspective

Standing on the edge of a mountain range and looking into the distance - as far as the mountains allow - the mountaintops in the foreground still appear in clear and rich colors, whereas the mountains in the background lose color with increasing distance, fading away in a milky white or light blue, depending on the percentage of humidity and the respective position of the sun, i.e. the time of day.

Old computer animations are often characterized by a certain artificiality, which is achieved by deliberately neglecting all color perspective. Atmospherical effects like the decrease in color richness are necessary, however, to make the 3D representation appear convincing and believable.

Especially in landscape representations the deliberate use of color perspective is not only a means for representing nature convincingly, but also a suitable tool for adding depth to a scene.

This effect can be achieved relatively easily, because most 3D programs offer their own fog effects. Some programs specialized in exterior representations go as far as relieving the user of this task by incorporating an automatic color perspective.

Normally this loss of color is achieved by the deliberate addition of white "noise" to the background of a scene.
The further the object is away from the camera, the more dense is noise filter covering the scene is and the weaker and lighter the colors appear.

**Fig. 152.** A landscape characterized by loss of color richness and contrast

In this respect it has to be kept in mind that the atmospherically loss of color will vary depending on the time of day, the temperature, and the humidity. On a sunny day in summer, color perspective will appear with a bluish hue, and at sunset on an autumn evening, it will appear more reddish.

It is difficult to represent this specific color effect in black and white; nevertheless, the illustration gives an idea of the effect described. While in the foreground the contours of the landscape are sharply defined, they disappear towards the horizon, in a diffuse haze getting denser with increasing distance to the camera. At the same time, the colors fade until only a milky blue remains on the horizon.

## Mist and Fog

In most 3D programs mist and fog are used in three different ways:

**Standard fog:** Here, the entire scene is covered by a filter. You can use opacity-maps for density or environment maps for the color of the fog.
**Layered fog:** Between specific upper and lower boundaries the layer of fog gets more dense or thins out. Rising ground fog is simulated, for example, with layered fog.
**Volume fog:** Here a fog is generated within a specific 3-dimensional space that does not have a constant density.

While the first two types of fog are generated by a filter applied to the finished picture, and are defined as a variable transparent layer between the camera and the scene, the third type is a 3-dimensional effect, which can be spatially described within defined boundaries by a so-called Gizmo. In a real-life situation this will look more or less like this:

### Fog as a Background

When fog is used for the background of a scene, a picture filter with a white noise is placed over the background area, thus achieving a fading effect on the colors.
Used unobtrusively, such a fog will simulate the effect of natural color loss. As soon as the fog is more dominant, it will become a design element for supporting the desired mood of the scene.

fog background on
the background of the picture
is influenced by fog

fog background off
the background of the picture
is not influenced by fog

**Fig. 153.** Fog and its influence on the background of the pictures

Fog as an atmospheric scenic background for supporting color perspective can either influence the geometry of a scene, or the background (e.g. background picture).
Fog with transparency information can also be used as a delicate cloud formation.

## Fog Density

As a rule, the increase of the density of the fog can be controlled either by a linear or by an exponential function.

the fog increases towards the horizon in a linear manner, near 0% far 100%

the fog increases exponentially towards the horizon, near 0% far 100%

**Fig. 154.** Linear or exponential increase of fog density

Certain percentages of the fog can usually be allocated to the foreground and the background of a picture. In this way, the fog density can be specified according to the requirements of a visualization.

near 60%, far 0%

near 0%, far 60%

**Fig. 155.** Values of fog density for the area in front and in the distance

### hint Transparent Materials

*Caution is advisable when using fog and transparent materials, or materials which make use of an active opacity map. In the example the road is "softened" at the edges by a gradient map with additional noise. By adding the fog the opacity map is ignored which can result in an error in the representation. The problem can be solved by separate rendering of the components. The results in the form of several layers can then be joined again, in a Video-Post program, such as Combustion or After Effects,- most large 3D packages generally include a complete Video-Post package ( in 3ds max under Rendering – Video post ...).*

## Layered Fog

With the previously presented fog, the effect of color perspective for areas in the vicinity as well as in the distance can be extremely well simulated. Another popular effect, used quite often, is the fog wavering just above the ground, in the early hours of the morning.
This phenomenon is called layered fog, which, instead of being defined from near to far, is now defined from top to bottom (or vice versa).

Bottom 1 Top 8, Falloff Bottom

Bottom 0 Top 0,5, Falloff Top

**Fig. 156.** Layered fog in different thicknesses. The falloff on the left passes towards the top, on the right towards the bottom

The Falloff is the parameter, which defines the transition from the layer of fog to the clear air on top or at the bottom.

The units are taken over from the pre-settings of the file. As the scene has been drawn up in meter units, 0,5 units are equal to 50 cm.

By changing the course of the fog from top to bottom, an excellent simulation of a low stratus can be done.

By additionally adding effects like horizon noise, the horizontal line, if visible at all, can be slightly blurred (right picture). Especially in views across large bodies of water, this effect can be very helpful for avoiding sharp edges/contours on the horizon.

**Fig. 157.** Ocean view with a very slight horizon noise

As a rule, fog and mist can, of course, also be animated.

## Volume Fog

In contrast to the previously mentioned types of fog, by using volume fog, as is indicated by the name, a 3-dimensional effect of fog, mist or even clouds can be generated.

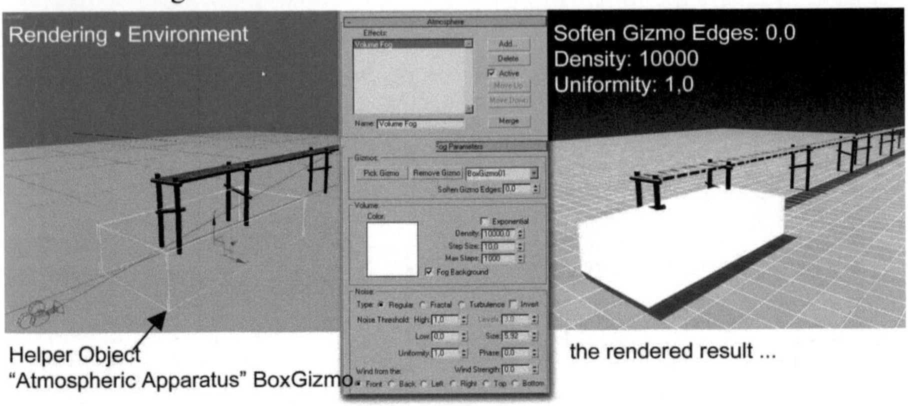


**Fig. 158.** Volume fog with very sharp edges for demonstrating the effect. The "BoxGizmo" serves as a limitation of the extension of the fog

For generating a volume fog, the rendering effect alone is not sufficient. In this case, the boundary constraint of the fog inside the scene has to be defined. This dummy can take on the form of a box, a sphere, or a cylinder, depending on the use. Within this helper object the density of the fog, the noise distribution, and the type of the falloff area is then defined.

The upper part of the illustration shows the behavior of the volume fog when all "soft" parameters are deactivated. The fog is so dense that it fills the sharp-edged dummy completely. It is also obvious that the fog behaves like a 3-dimensional object, which is able to cover up other objects in the scene and has a shadow as well.

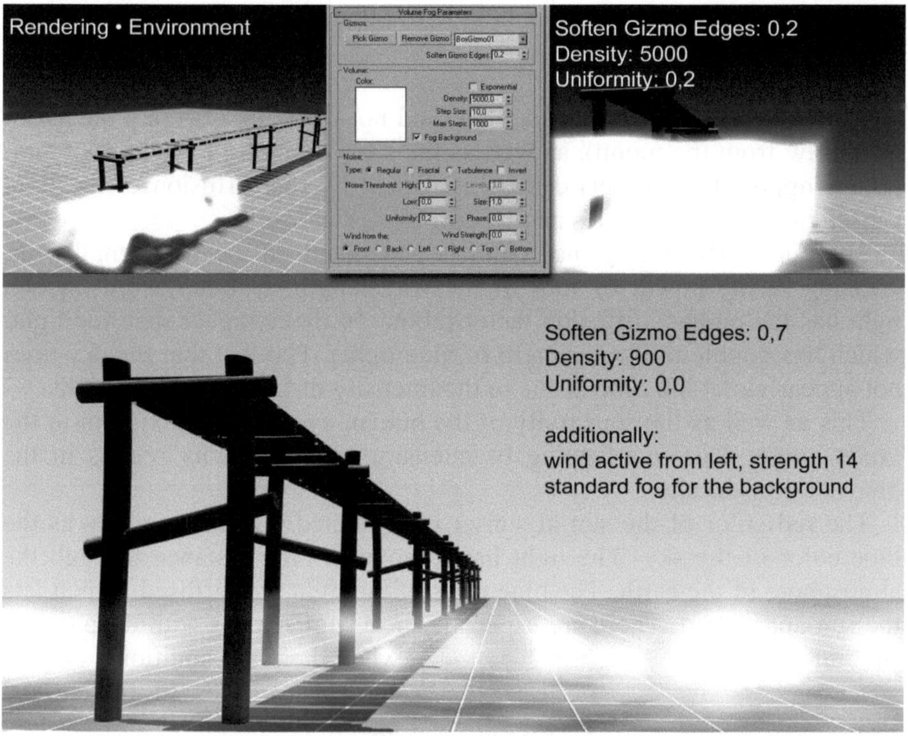


**Fig. 159.** Soft edges and reduction of thickness ensure a suitable appearance

The settings of the respective parameters are subject, like so much else, above all to personal experience. It may be helpful to use the system units as a measurement for size. The parameter values, as a rule, correlate to the units of the file (as always, there are exceptions). Therefore, one should always keep an eye on the units applied.

## Sky

However, why is the sky actually blue?

This question, normally asked by children, has kept the authors quite busy, because even some great scientific minds are not sure about the answer. One answer following the current scientific opinion, seeming plausible and understandable, can be found on the homepage of Hans Schremmer[1]. Here, you will also find a couple of other phenomena of the sky discussed in depth.

The fact that the sky appears blue (and not black like the sky the astronauts saw from the moon), as well as the slow dimming of light after sunset (as opposed to abrupt) can be explained by the diffusion of the light caused by the air molecules and by particles of aerosol.

The pure diffusion by molecules is dependent on the wavelengths, increasing to the power of four of the wavelength; in other words, violet light has the highest diffusion factor (about 16 times higher than red light, which has double the wavelength of blue light). The fact that the sky does not appear violet but blue is due to the intensity distribution of sunlight.

This as well as the sensitivity of the human eye has its maximum in the green range. This overlapping of intensity and sensitivity results in the blue color of the sky.

The red color of the sun at sunset is explained in the same way as the blue color of the sky. The light has to cover a long distance through the atmosphere of the earth. The blue parts of the light are diffused, which lets the sun appear red. There are especially beautiful sunsets with deep colors when there are additional dust or aerosol particles in the atmosphere.

**Fig. 160.** Sky.JPG from the collection of 3ds max

Who does not know the standard cloud backgrounds, which always seem to be providing the 3D background in an incredible number of architectural or landscape visualizations? Most 3D programs have a certain repertoire of sky pictures, which normally is quite sufficient. But when certain cloud formations have been seen too often, they are likely to become boring.
One cannot always avoid falling back on a “quick” and ”instant” background, especially when there is a lot of pressure to finish a project.

However, with a little extra effort, it is possible to achieve convincing picture backgrounds, even using existing materials.

[1] http://www.schremmer.de

The sky above us has a lot more to offer than a few repetitive background pictures. Therefore, it is worth the while to take a closer look at the possibilities and methods of generating and manipulating the firmament.

Two types of sky can be distinguished, namely the use of a background picture, and a procedurally generated sky (similar to the generation of materials):

**Background picture** – any photograph or painted picture
**Procedurally generated sky** – generating a sky is done by the respective algorithms

It is, of course, the mixture of these types of representation that actually leads to a large variety of further details.

## Background Picture

As long as some basic aspects are observed, the use of a background picture will provide the quickest result.
Background pictures can be private photographs, or picture materials loaded down from innumerable Internet platforms, either free of charge, or at a certain cost.

By entering the terms “background pictures and/or sky” the current search engines will quickly deliver a multitude of suitable pictures, often freely available.

When using such a background picture for a landscape visualization, one has to keep in mind that:

The resolution is sufficiently high to provide the scene with enough background (the dimensions of the picture should correspond at least to the desired rendering size)
The picture should be tileable, if the scene has to be provided with a celestial sphere or a spherical environment and there will be large camera movements/ pan /shot/tilt
The setting of the environment map (see fig. 162) is fitted to the use in the scene, and of course
The picture is suitable for the design of the scene

## Background Picture for a Still

It takes less effort to provide the still frame of a landscape picture with a suitable background than to provide an animated sequence with a background.

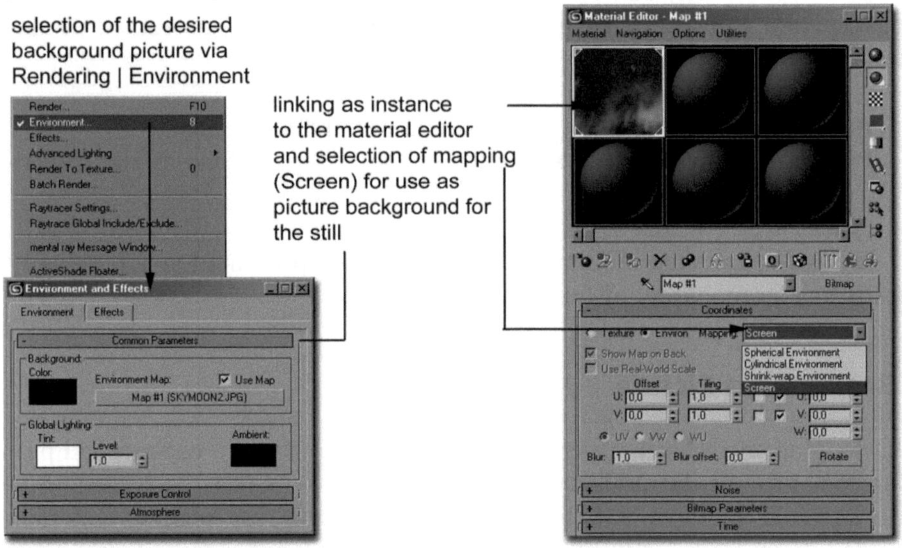


**Fig. 161.** Background picture in the rendering settings *ENVIRONMENT AND EFFECTS* • *ENVIRONMENT MAP*

## Background Picture for Animations

For an animated sequence with a variable camera position the following procedure is recommended:

After generating a hemisphere (the closed half of a sphere), which covers the entire scene, and the normal vectors of which are directed inwards, i.e. into the hemisphere, a suitable picture is chosen. Here, one should keep in mind that the picture has to be tileable to prevent a collision of the picture edges.

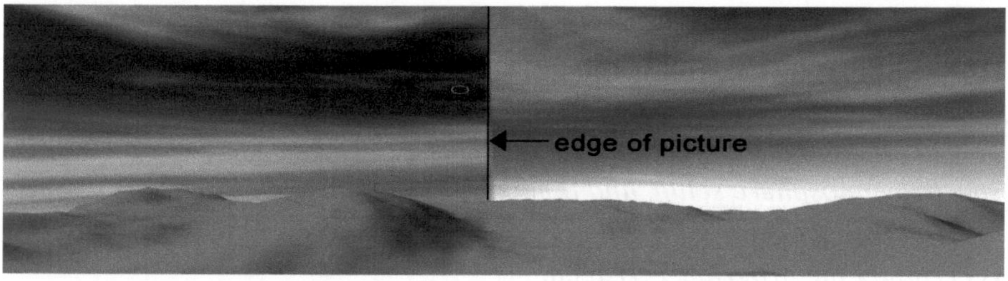


**Fig. 162.** In non-tileable pictures the edges will collide sharply

Alternatives for avoiding this are either to do without focusing the camera towards the edge, or covering the edge by adding specific objects, such as a tree, or a building.

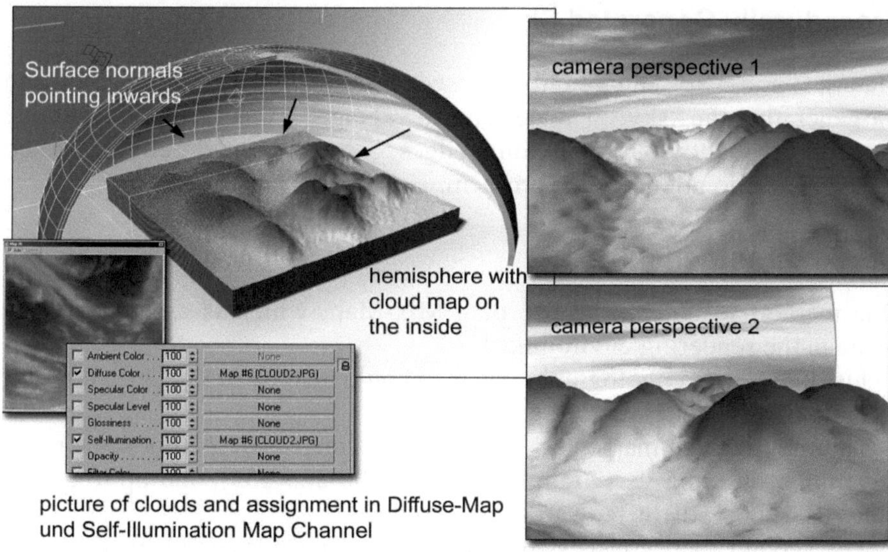


**Fig. 163.** Generating a hemisphere with a texture directed inwards, to represent the firmament for later animation

### Tip Special Software for Backgrounds

*Some manufacturers specialize in the generation of landscapes and the respective atmosphere setup, and it makes sense to take a closer look at one or two other tools for generating background pictures. One of these tools is certainly the program Terragen.*

*This terrain visualizing tool can be used free of charge. However, there are some limitations concerning the size of the picture to be rendered.*[2]

## Procedurally Generated Sky

Materials and maps cannot only be used for covering any surface, but also for generating the "sky". The decisive advantage of this method is that one needs not pay attention to colliding edges, because procedurally generated backgrounds do not have this problem.

### *Mix-Map for the Sky*

One method would be the use of a Mix-Map, which is assigned to the scene in the rendering environment.

Just like using a photograph, this map could also be in the Diffuse Color channel of any material.

The first of three maps used in Mix-Maps consists of a color scale ranging from white to a very light blue (RGB 200 225 255), and up to a dark blue (RGB 60 110 180) towards the top edge.

The second map consists of a noise-map, the horizontal distortion of which generates a cloud-like appearance.

The third map consists of a pure black-and-white color scale. In the mixing channel of the mixing maps, the scale generates the information for blending the two materials. It provides a mask, in other words, the white areas being opaque, and the black areas being transparent.

[2] http://www.planetside.co.uk

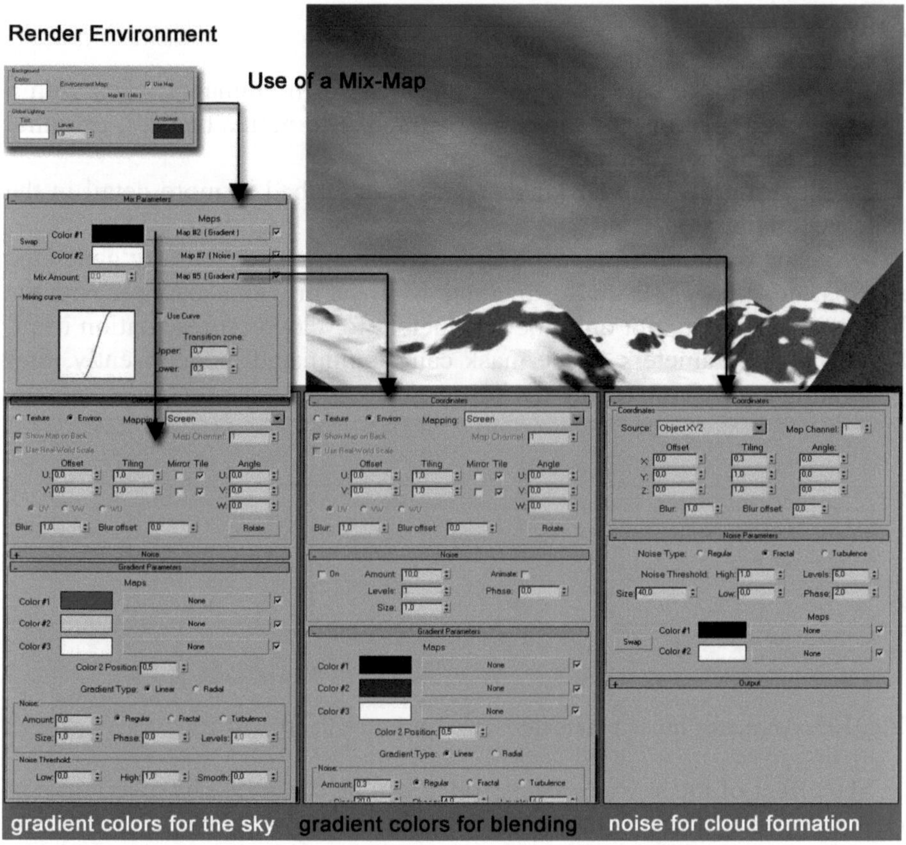


**Fig. 164.** By using a Mix-Map two maps are blended into each other

By adding an additional noise to this procedure, the background appears more irregular and more natural.
Not only the celestial body can be created quickly and pleasingly with the help of a procedural map, but also the coloring. The color of the sky change during the course of the day, from shades of red at dawn towards a deep blue in the middle of the day up to shades of red with hues of violet at dusk. Various lighting moods can be enhanced by carefully choosing the color of the sky.
A cloudy day with a high percentage of grey, or the strange mood just before a storm with some yellowish light are just two examples for this. A sunny day with a couple of cumulus clouds is not always the right choice for the successful composition of a scene; sometimes it pays to explore the possibilities of the weather in more depth and detail.

## Animating the Sky

Not only when the camera is moving, but also in animated takes with a fixed camera position, it can be very attractive to notice the sky, or rather the cloud formations, in motion.
The generation of 3-dimensional clouds is described in more detail in the following paragraph.
When using a mere map for the sky, certainly the procedurally generated map has advantages. With this method, as in the previous example with the help of a Mix-Map, not only the parameters for cloud representation itself, but also the parameters of the mask can be animated independently, thus generating a very convincing background.

**Fig. 165.** Animated noise parameter "size" at frame 0, 50, and 100

The effect of the sky as a background can be supported by a certain percentage of self-illumination. Apart from the animation of cloud formations, the color variation during the course of the day can very easily be achieved via the animation parameters of a procedural map.

Generally, such results can also be achieved by additionally using picture-editing programs for background pictures, but the mathematically controlled method of the procedure maps offers considerably more possibilities.

## Clouds

Actually, the simplest method of generating clouds has already been presented when we dealt with the sky. A 2-dimensional map is either used as scene background, or mapped as a material on to an hemi-sphere or another background object. This way, with the help of a noise filter, procedural cloud backgrounds can be quickly generated, and even animated without too much of an effort. The alternative to this is picture editing and "painting", with pen and brush, a cloud background in Photoshop or a similar picture-editing program. Incidentally, a very good suggestion regarding the subject of various clouds can be found in the "Karlsruhe" Cloud Atlas under: http://www.wolkenatlas.de.

But what, if this background effect, even if supported by fog, is not sufficient? What is to be done when clouds with volume are necessary and the camera has to fly through?

A few typical requirements could be:

The clouds have to appear 3-dimensional
A fly through has to be possible
The clouds have to cast shadows
An animation should be possible

One suitable method is the use of the previously explained volumetric fog effects. It is recommended to spend some time playing around with the parameters. The success will come quickly and without too much effort. However, unfortunately the quality of such clouds generated by fog parameters is not perfect. Mostly, they are lacking in plasticity; thus they may be quite sufficient for a cloud scenario in motion, with animated shadows, but not for much more.

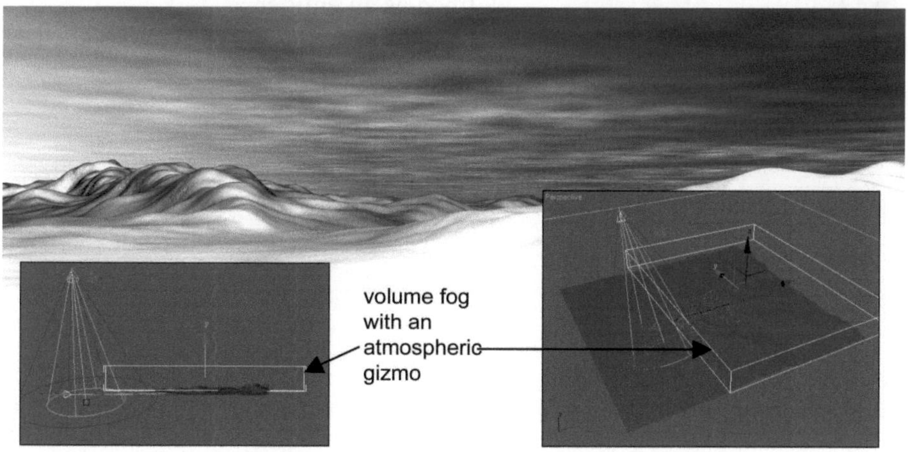


**Fig. 166.** Animated cloud background with volume fog

A further method would be the use of a suitable plug-in for clouds and smoke, such as Pyrocluster[3], Afterburn[4] or others. Nevertheless, this means

[3] http://www.cebas.com - Cebas
[4] http://www.afterworks.com – Sitni Sati

extra cost, and therefore is not always a suitable alternative. However, clouds can be done very well with particle systems.

### Tip Particle System in Max

*In 3ds max the Particle System is called up under CREATE, GEOMETRY, PARTICLE SYSTEMS.*

In an example, this could look more or less like this:
As a basis, you need

A Particle System with the possibility to instance geometries
A geometry for referencing
A suitable background

**Fig. 167.** Front view of particles

The pre-defined particle system *PCLOUD* includes, similar to the Gizmos on volume fog, several instant dummies, within which the particles can be generated. Alternatively, any geometry can be defined as an emitter.

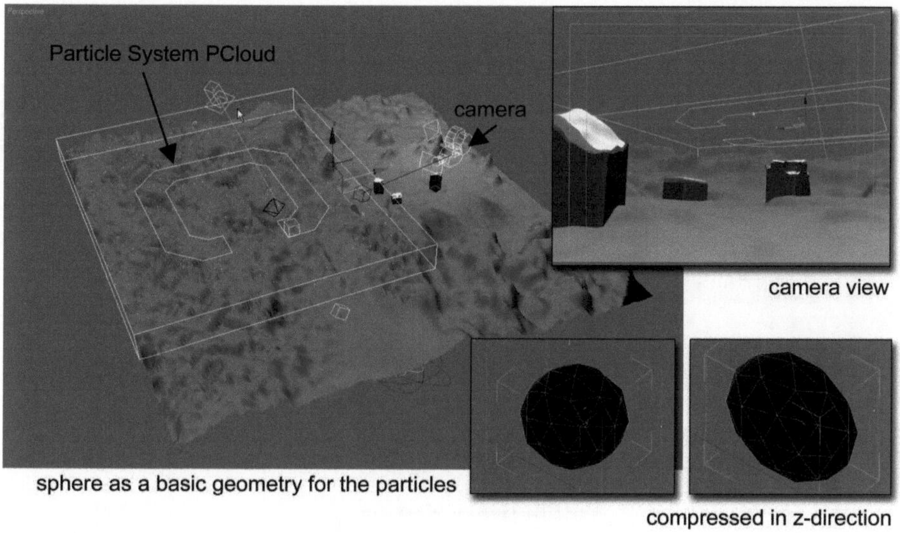


**Fig. 168.** Particle system *PCLOUD* for cloud formation

A referenced geometry - a sphere is created. It is squashed towards Z. This offers the advantage that later the clouds will not appear too much like "bubbles".

If the sphere is chosen as a particle, and the number increased so that the impression of a cloudbank is created, the resulting picture will look more or less like in the illustration above (left). Now the problem actually only lies in finding a suitable cloud material which gives the "clouds" a more cloud-like appearance.

The first possibility of getting a good result is the use of a noise map. Firstly, the coloring of the cloud has to be defined. This is done with a noise map in the diffuse color channel of a standard material.

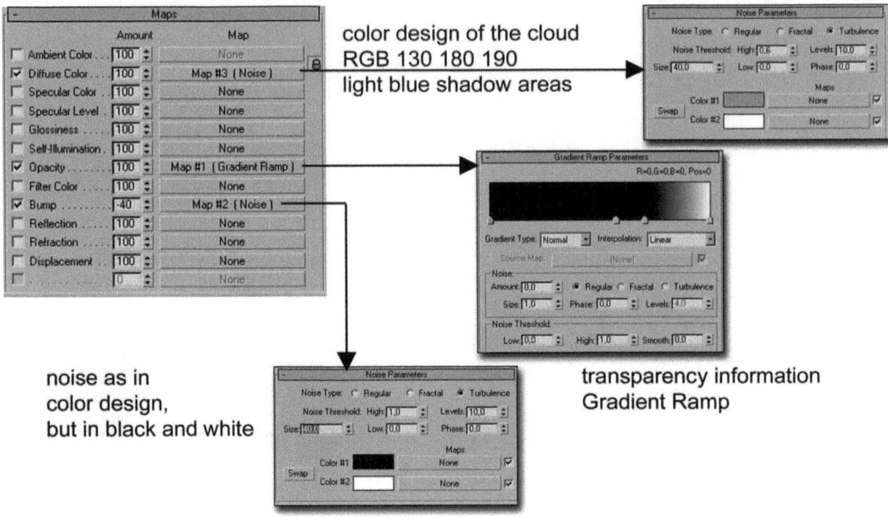


**Fig. 169.** Material for clouds

The *NOISE* map in the *DIFFUSE* channel is responsible for the color design of the cloud. Here, light areas (white) and "shadow" areas of the cloud (light blue) are assigned.

By copying the *NOISE* map into the *BUMP* channel and changing the shade of blue into black, and assigning a negative sign to the *BUMP* (relief) value (-40), this second *NOISE* will provide the cloud with the necessary plasticity.

Because the values are dependent on the units of the scene, the numerical values given here are to be treated with caution.

What is still missing now is the transparency information of the cloud. This is achieved by the map *GRADIENT RAMP*.

In order to prevent the cloud from casting too much shadow, it is recommended to deactivate the attribute "conserve shadow". This way the cloud object is still able to cast a shadow, but there will be no shadow between the separate cloud areas.

The clouds generated in this way cast a shadow, appear 3-dimensional and can be quickly and efficiently animated, like all particle systems.

Different moods can be adapted quite quickly to any situation by changing color, behavior and design of the particle-based clouds.

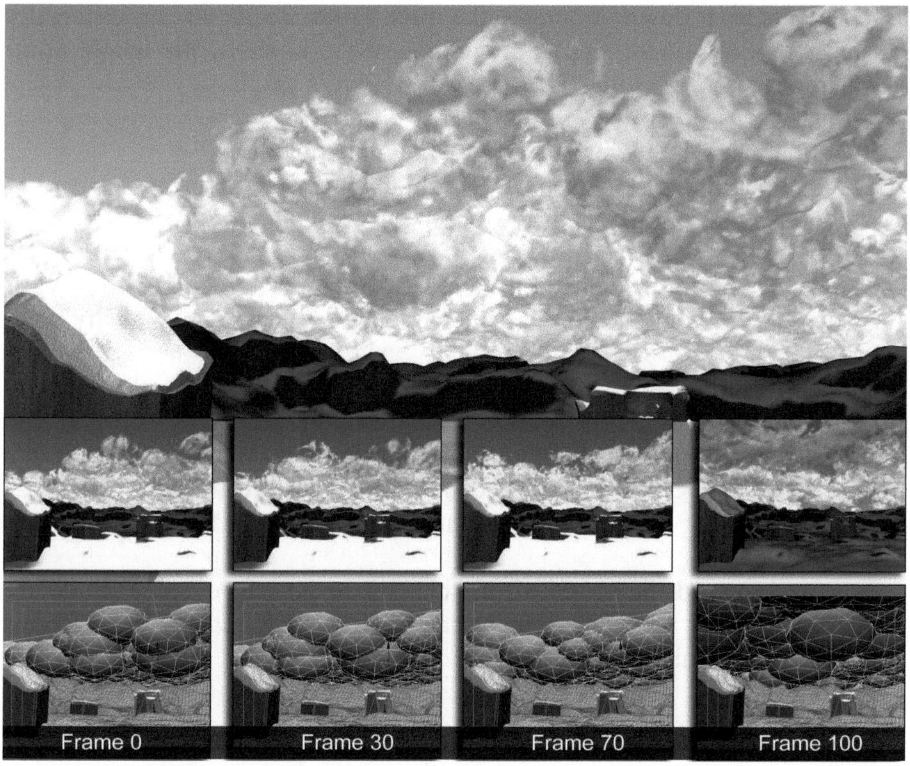


**Fig. 170.** The finished particle clouds

## Rainmaker

Some noise and a little soft focus in Photoshop, and the rain is done for a still frame. To go into more detail only makes sense if rain has to be used in an animation. The basic requirements for rain in an animation can be reduced to the following points:

A simple particle system for the 3D scene
Sufficient particles and a suitable texture
A little Motion Blur

… and the precipitation event is finished.

But when getting down to details, the good times are soon over, especially as far as "quick" and "simple" are concerned – in short, the whole affair starts to get really exciting. The creation of rain is tough.

What happens at the very moment when it starts to rain?
What about all the rain drops that are bouncing off the surface?
How does rain interact with the effects of wind and gravity?

Rain is generally influenced by wind. The wind often comes from various, changing directions, and has the tendency to change its intensity quite often.

On hitting a surface, a raindrop is either absorbed and reflected only marginally, or it is nearly totally reflected after hitting the ground and splitting into innumerable smaller units of raindrops. When a raindrop hits a water surface, very interesting concentric ripples are formed, creating a lively chaos after the reflected minute drops have hit the water surface again.

It would exceed the scope of this chapter to go into every little detail; however, a closer look needs to be taken at some fundamental questions surrounding the subject of rain.

## The simple Variation

Let's say that it has been raining for quite a while in the virtual space. Therefore, all objects in the scene are wet already. Thus, firstly we need to represent the rain, and the secondly to represent the wet surfaces.

Just as when dealing with plants and clouds, particle systems come to mind. Particle systems can use simple basic primitives as well as referenced geometries as particles; they can react to forces and simulate collision behavior.

Going even further, there are rule-based particle systems to which can be assigned a specific behavior at a specific time, or by interaction with other objects or events.

As a rule, for a simple rain shower a simple particle system is sufficient. In the present case the particle system *SUPER SPRAY* is used to generate just such a rain.

In the scene, a particle system *SUPER SPRAY* is generated above the flowerpots. The number of drops is 10.000, their size is one unit (1), the speed is 10, and the variation of generation amounts is 2. The particles have a life span of 30 frames dissolving into nothing in the end.

These settings refer to the file at hand, in 3ds max. As a general rule, it is recommended to first test the generation of particles and the number of possible parameters in a simple example, in order to get familiar with this tool.

**Fig. 171.** Installing a particle system

At frame 20, it is already raining cats and dogs and it can be noticed that the drops are hitting right through the floor on which the pots are standing.

In other words, the effect is already quite nice, but what is missing is the reflection of the drops on the floor, - those little droplets of dancing and bouncing water which can look rather funny, depending on the incidence of the light.

**Fig. 172.** Particle system in action

To prevent the drops from penetrating the floor, the use of a so-called deflector is needed. A deflector takes care of the interaction of a particle system with a surface or any geometry and describes the behavior of the impacting particles, i.e. it fits the reflection to the particle hitting the deflector.

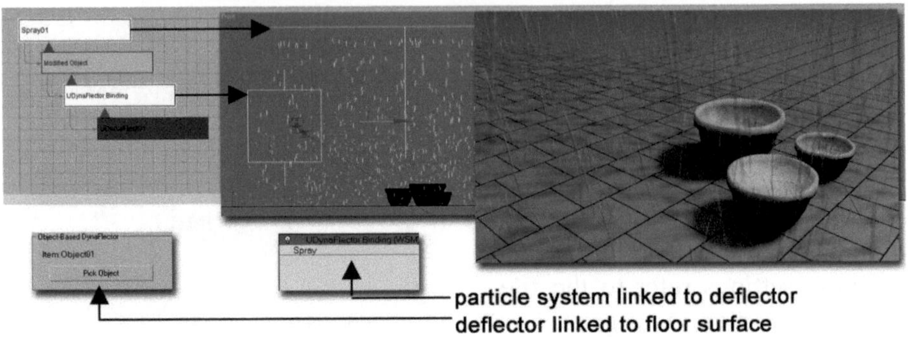


**Fig. 173.** Particle system reacting to the deflector and drops bouncing off the floor

A second deflector sees to it that the pots, too, are defined as a resistance, and that there are no unwanted penetrations.

Because the whole scene has to be wet, it is necessary to change the materials into a "wet" state.

In order to achieve this, the reflection values of the various materials are increased, and an additional Raytrace Map for the mirror effect reflection is assigned to the floor.

**Fig. 174.** The materials were fitted with reflection and wetness

A further phenomenon, which is likely to drive any visualizer to despair, is the stage when the rain has just started. The floor is as dry as dust, and one notices those wet spots beginning to spread out across the floor.

## What happens when it starts to rain?

As always, there are various ways and means leading to a result. One possibility would be, for instance, the use of an event-based particle system, like Particle Flow. Although very powerful, this method is unfortunately also full of obstacles and a stony path.

Another possibility is the use of the material editor.

So, what happens when raindrops fall on dry ground?

They impact, leaving small craters, splashing and distributing some surplus water into the air, -but we do not really observe these effects. All we notice is the fact that suddenly and quickly; "dark" spots start to form. There are more and more of them, and they do not disappear, but soon cover the entire floor.

If this actual information is abstracted to a simple model, two materials will be quite sufficient for a 3D environment, - namely one material for the original state (floor dry), and one material for the later state (floor wet).

Taking into account that the floor is at a certain distance from the viewer, or that the point of interest is directed to a specific event in the scene, which would make the falling rain important but not dominant, then a possible solution could look more or less like this:

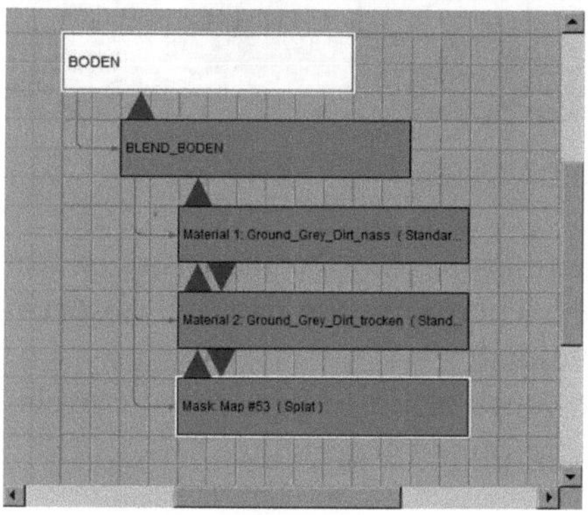


**Fig. 175.** Blend Material

A material for the floor is generated, - in the present case a standard material with a bitmap in the DIFFUSE channel, and the same bitmap in the BUMP channel.

When this material is copied and the settings of the reflection (e.g. Raytrace-Map) and the highlights are changed, the material will appear wet.

Via a mask, the two materials are blended into each other and the event of starting rain is achieved.

### hint Blend Material

*This whole procedure can be imagined like this: Two pictures are put on top of each other (like two transparencies). Additionally, a "mask" is used to allow the information of the picture on top to "pass through" into the picture on the bottom. A mask works like an Alpha Channel with black and white information for the representation of transparency (white = opaque; black = transparent). A mask can be a picture, a movie, an animated sequence, or any procedural map.*

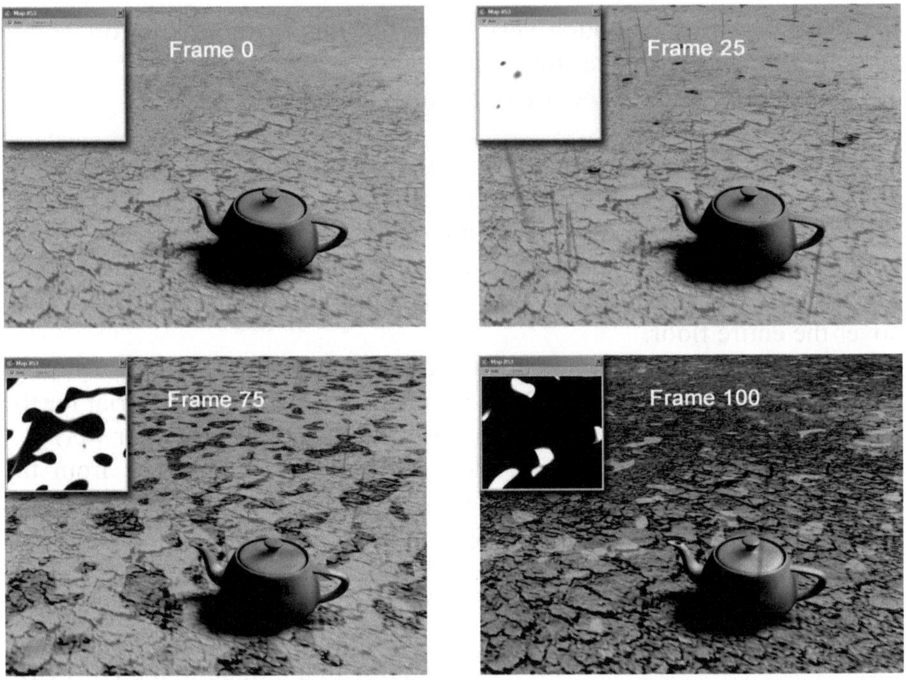


**Fig. 176.** BLEND material with animated SPLAT map

The problem here is that the falling raindrops and the increasingly wet ground, do not correlate with each other. Therefore, the detailed observation of a single falling drop should be avoided.

It is, of course, more elegant but also more effort to generate a correlation between the drops (particles) and the ground (material).

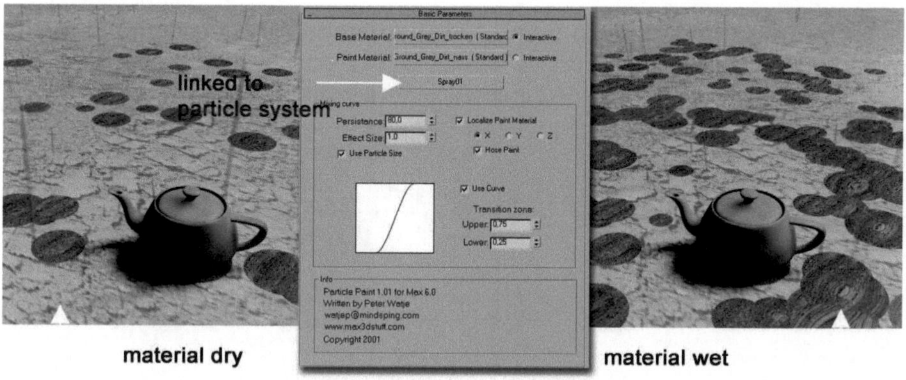


**Fig. 177.** Special material by Peter Watje [5], which reacts to falling particles. Here an automatic blend to a second material is generated on the spot, which has been hit by a particle

## Snow

Falling snow is certainly not just an interesting possibility to present a landscape in a winter mood; it brings back memories and creates moods. Maybe you remember some childhood days, when you were staring out of the classroom window instead of following the lesson, fascinated by the strange pattern of falling snowflakes and completely forgetting what was going on in class.

The consequences of not paying attention in class were usually unpleasant. Snow, especially when falling, behaves similarly to fire, - one can stare at it for hours and completely forget the time.

The manner in which snowflakes fall down is especially interesting. They do not just fall straight down. No, they are blown away by the wind, forming whirls, even partly moving upwards, before slowly and majestically finally reaching the ground. If the ground is still too warm, they melt immediately upon impact, if it is cold enough the freshly fallen snow forms whirls even on the ground.

Although rain falls at such a high speed that the ensuing motion fuzziness will actually always lead to good results, the situation is altogether different with falling snow. The flakes sway near the camera, and there is sufficient opportunity to regard a single snowflake in detail.

---

[5] Free Plugin, Download under http://www.maxplugins.de

Therefore it is important to give special attention to the generation of the material to be used.

**Fig. 178.** Particle system for generating snow with the help of an instanced geometry

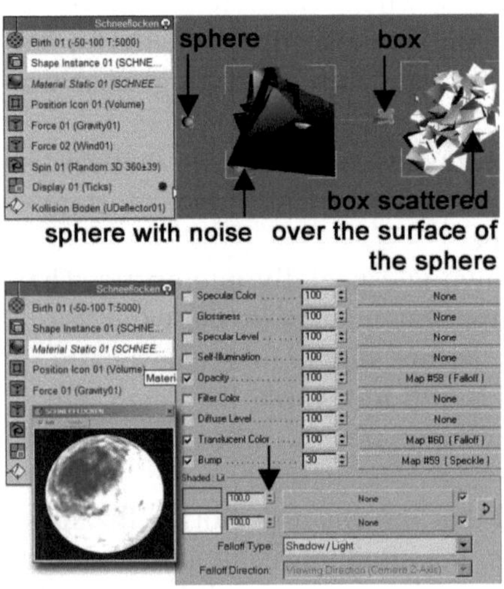


**Fig. 179.** Snowflake and material with Translucent Shader

In the picture above left the particle system *PF SOURCE* was generated.

As an alternative the particle system *BLIZZARD* could also have been used.

However, *PF SOURCE* offers a large variety of parameter settings and enables more interference by the designer.

Among others the particles in the example have been linked to an outside force, namely gravitation, and wind. A sphere, which has been distorted via noise, and the polygons of which have been edited additionally, serves as a particle element.

### Hinweis Translucent Shader

*Translucent Shaders are also used for the simulation of human skin.*

The material used is a so-called Translucent Shader, which, like in a real snowflake, transports light within the surface, and thus, although not appearing totally transparent, keeps its plasticity.
The settings for the material were as follows:

**Shader:** Translucent Shader
**Diffuse Color:** White
**Opacity Map:** Falloff
**Bump Map:** Speckle

This suggestion for generating snow has been taken from the book by Pete Draper “Deconstructing the elements with 3ds max 6”.

## Summary

We have discussed the atmosphere, - atmosphere as a design element for moods, and how the atmosphere can be generated in a 3D visualization, especially in landscape visualization.
A quick course touching on the predominant atmospheric effects served as the basis for design, giving ideas and incentives for experiments. Particularly when dealing with special effects, of which most atmospheric phenomena are a part, it is difficult to look at the subject matter without considering the respective tools. The attempt to discuss this subject in general terms is especially difficult. Some readers may have found this chapter too program-specific, - but in our opinion this was impossible to avoid.
Firstly, terms like color perspective and how this is manifested in nature, were dealt with. Mist and fog can be generated with 3D tools via so-called rendering effects when post-processing the finished picture. The generation of mist and fog can roughly be divided into pure filter effects (fog), and volumetric effects (volume fog).
The effect of color perspective can be simulated by such filters.
When generating a “sky” as a picture background, generally a map is used. This can be a normal photograph, any digital picture, or a procedural map, i.e. based on special algorithms.

Picture backgrounds available as a bitmap have the advantage of quick implementation in still frames. In animated sequences, their tile ability has to be kept in mind.

Depending on the purpose of the visualization, a secondary animation of the sky is recommended, as can be achieved for instance by moving clouds. In animated sequences, a procedurally generated sky is more flexible and more versatile than a "pure" picture based on pixels.

On occasion, it makes sense to mix both methods for obtaining the desired result.

The use of volumetric clouds will lead to an "improved" sky as a picture background. These can be generated quickly and with some elegance with the help of particle systems, and can be animated in a variety of ways.

Rain will fall from the clouds, and here too, the use of a particle system for generating rain was proposed. A further important aspect concerning rain is the transition from dry to wet. One possibility of representing this process is the use of suitable materials, like for instance the Blend Material.

Snow, just like rain, can be generated quickly with the help of a particle system. However, as opposed to the fast falling precipitation form of rain, suitable reference geometry has to be generated and fitted with a suitable material to serve as a snowflake.

What remains are many unanswered questions, heaps of ideas and, hopefully, lots of fun when practicing.

# Water

*Out of the rain and into the gutter* [1]*.*
*While rain is a phenomenon belonging to the atmospheric effects, the effect of the gutter (or eaves) undoubtedly belongs to the field of animated water. There was a time when there were no gutters. At this time, during a heavy rain shower, nothing worse could happen to a person than to move out of the rain and under that part of the roof where the water that had first hit the roof came down like a torrent. The present chapter will deal, among others, with this subject. The main topic is water and the interesting forms in which it occurs in landscape visualizations.*

**Fig. 180.** Decorative spray of water in front of the Bellagio Hotel (Photo: J. Kieferle)

Bellagio: Have you ever been to Las Vegas? No? You should definitely go there, but not to lose your money in a casino, but to study the special effects developed by the company Water Entertainment Technologies (WET). WET is well known for its water sculptures based on the physical phenomenon of laminary flow, i.e. the steady, continuous, non-turbulent movement of single particles. Under high pressure, water is pushed through specially constructed jets which prevent the water from taking in air bubbles; this way a glasslike body of water is created which will disappear into special openings in another part of the fountain. When playing with the hosepipe in the garden, one can observe similar effects.
Now….
When thinking of water and the effort that has to be invested in the hydronumerical field in order to achieve, in various models, a flux precision in the realm of decimeters, it seems preposterous to even think about an acceptable simulation of water. On the other hand, when watching movies like "The Day after Tomorrow", or "The Storm" and realizing that only

[1] Eaves or Gutter - Eaves – the projecting edge of the roof. From the rain into the gutter – German saying which means from bad to worse or "out of the frying pan into the fire"

visualization tools were used here, one can safely forget the numerical approach, and rely instead on a good eye. Because in the end a good observation of nature and the attempt to represent this with the means of an artist will be quite sufficient here. Although one should not forget that there are entire departments in some companies dealing with special effects, are devoting their energies to the generation of raindrops, it is obvious that convincing results in the representation of water can be achieved even without numerical approaches.
And that is what counts here - some of the basics for representing water in various forms. We have tried to give a summary of the most important manifestations of water within the visualization of landscapes. These manifestations are:

**Water surfaces** - lakes, ponds, and puddles
**Streaming water** - rivers, streams/brooks, canals
**Tumbling/falling water** - waterfalls, precipices, outlets/drains
**Border and transition areas** - shores and riverbanks

Each of these manifestations will be examined with respect to its appearance and to its transposition into a 3D visualization.

In addition we will look at the important role of the specific characteristics of water, namely reflection and refraction, the matter of its different physical states, and of course the generation of suitable visualization materials.

But first let us look at a few basic facts about water.

## State of Aggregation?

Water certainly is one of the most interesting "elements", because no other matter acts so "naturally" strange. Under normal conditions water is a liquid, at high temperatures it turns to vapor, and below zero it is extremely painful to be hit on the head by a frozen piece of water. Water is the only known matter in nature appearing in all three physical states, i.e. liquid, solid, and as a vapor.

**Fig. 181.** Three physical states of water in one scene

Not only do we humans consist to a high percentage of water, water is also the matter, which dominates our lives. The melting as well as the boiling point of water is of such importance to us that they have become fixed quantities (0°C and 100°C) on temperature scales.

## Further Specific Characteristics

The anomaly of water is responsible for the fact that, under normal pressure and at a temperature of 4°C, it has its highest density of 1000 grams per cubic centimeter. When the temperature goes down further, below 4°C, water will expand again, which is contrary to the usual behavior of any other material, which increases with increasing temperature. Who has never come across a situation where the radiator burst because somebody forgot to add antifreeze before the onset of winter?

Incidentally, Hannibal managed to build "roads" across the Alps by drilling holes into the rocks and filling them with water. The frost would then blow up the rocks.

But let us go on with those details that are important for visualization. The most important optical characteristics of water are its ability for reflection and its behavior of refraction. And these are built into the materials.

**Fig. 182.** Barcelona Pavilion – Pool with quiet water in the Barcelona Pavilion, 1929, Mies van der Rohe, Barcelona, Spain

The first thing that comes to mind, especially in computer animations, is water surfaces as clear as glass and highly reflective, mirroring a brilliant standard sky. But that is not all there is to it. Water surfaces captivate by their variety, they are never the same, and in real life water is very seldom as clear as glass.

There are various ways of representing water surfaces convincingly. Depending on the purpose, a simple blue surface may be quite sufficient in one case, while in another case it may be necessary to represent the surface of the water as realistically as possible.

## Water in Landscape Architecture

In the field of landscape architecture the component water is a fixed planning quantity, and its representation starts in most cases with a "classic plan".

All along, planners have had to represent water in a very special way.

The tools for the creation of convincing representation were pencils, watercolor, and, on occasion, airbrush. Nowadays, the possibilities of 3D visualization make this an attractive tool. The option of generating reflections automatically certainly is particularly attractive when dealing with water. One has to realize, however, that an attractive drawing will sometimes outdo a computer generated representation.

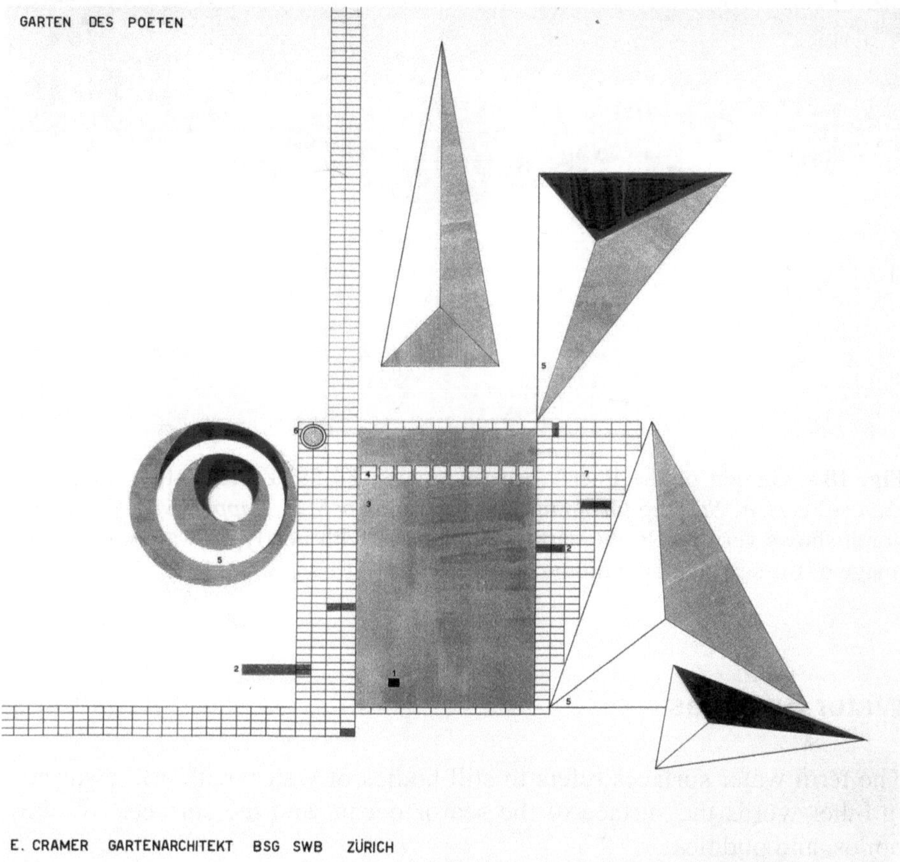


**Fig. 183.** Planning sketch for the "Garden of the Poet" - Ernst Cramer, Zürich [*Schweizerische Stiftung für Landschaftsarchitektur SLA, Rapperswil*]

„The garden was not so much a garden as a sculpture to walk through – abstract earth shapes independent of place, with sharp arises foreign to the nature of their material", ( Elizabeth Kassler: Modern Gardens and the Landscape, The Museum of Modern Art, N.Y.C., 1964, S. 56).

**Fig. 184.** Garden of the Poet - Ernst Cramer, G | 59, Zürich, after completion [*Schweizerische Stiftung für Landschaftsarchitektur SLA, Rapperswil*]. The photograph shows very nicely the dark water surface with nearly no waves, the mirror image of the sky and the building.

## Water Surfaces

The term water surfaces refers to still bodies of water, with a free surface, in other words the surface of the sea or ocean, and the surfaces of lakes, ponds, and puddles.

The color of the water surface is mainly dependent on two factors:

The color and composition of the sky and
The characteristics of the water.

The color and composition of the sky are reflected by the surface of the water. The reflection is not always completely symmetrical. Apart from other factors, waves and the loss of color intensity with increasing distance to the viewer have a strong influence on the coloring of the water.

**Fig. 185.** Water Surface

Another very important effect, which was discovered by the French physicist Augustin J. Fresnel[2], and which, in the meantime, has made its arrival into nearly all 3D programs, as a shader-effect, is the Fresnel effect

## Fresnel Effect

The so-called Fresnel Effect has to do with the varying strengths of reflection on a surface. These are dependant on the angle between the rays of sight and the object surface looked at.

In everyday language this can be explained like this: if you are standing on the shore of a lake or a riverbank and look towards the horizon, you will notice that although you can see the reflections on the surface of the water, it is nearly impossible to look beneath the surface. But when you look directly to the bottom you can see the ground, with no reflections in the way - provided, of course, that the water does not contain too much suspended matter, or that it is not disturbed by waves.

That is the Fresnel Effect!

[2] Augustin Jean Fresnel (* 10. May 1788 in Brogue (Eure); † 14. July, 1827 in Ville-d'Avray near Paris) was a French physicist and engineer who made important contributions to the establishment of the wave theory of light and optics

[3] Augustin Jean Fresnel (* 10. May 1788 in Brogue (Eure); † 14. July, 1827 in Ville-d'Avray near Paris) was a French physicist and engineer who made im-

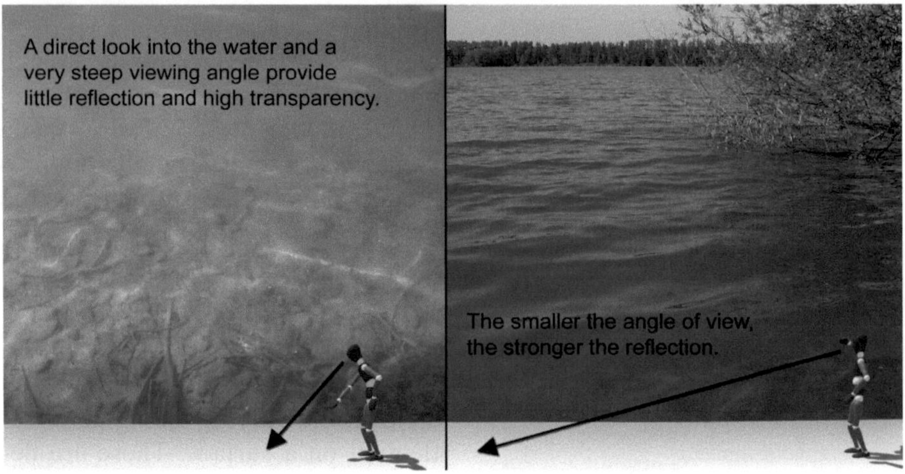


**Fig. 186.** The Fresnel Effect

## Waves on an Open Surface

An important aspect in the representation of an open water surface, be it the surface of a lake or of the sea, is above all the consideration of the waves. When studying waves in more detail, the movements on the surface of the water of large non-running bodies of water can be described as follows. Usually, surface waves are what we are dealing with. Special types of waves like those caused by a seaquake will not be discussed here.

There are waves that behave similarly but not identically, moving in similar directions. There are varying repetition frequencies and cycles when specific waves occur (if you are surfing you will know this waiting for the next suitable cycle of waves).

Looking even closer you will notice that the surface of the "big" waves can be subdivided into further small waves. Those small "ripples" are caused mostly by the wind.

When converting this reality into a 3D visualization, one will, as always, try to get by with a very simplified representation.

---

portant contributions to the establishment of the wave theory of light and optics.

## Example of a Water Surface

**Fig. 187.** A water surface with Noise and Glow Effect

A simple example for creating a representation of an open water surface like the ocean or a big lake could look like this:

### Generating a Plane

Firstly, the water surface is generated - a plane with very large dimensions offering a wide view onto the scene.

The units are defined in meters. Length and width are 500 each, subdivisions are 50 each in X and Y directions.

In the present case the *RENDER-MULTIPLIERS DENSITY* was set to 10. This way the number of segments is multiplied by factor 10.

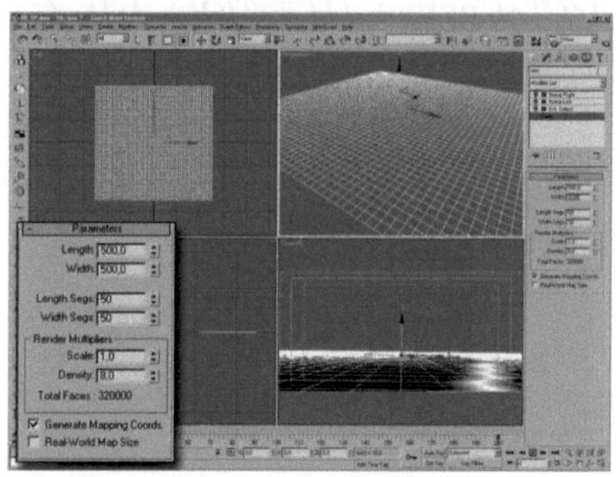


**Fig. 188.** Generating a Plane

### Tip "Vol. Select" instead of "Mesh Select" or "Poly Select"

*EDIT MESH, EDIT POLY, SELECT MESH, SELECT POLY all offer, just like VOL.SELECT, the possibility of editing sub-objects of a mesh either directly, or of using them as constraints for a modifier to be used later.*

*Using SELECT MESH or SELECT POLY the sub-elements, like Vertex, Edge, Face are marked via their sub-element index*[4]*. When refining the structure of the mesh later on by tessellation or deleting of elements, the selection of sub-objects will not tally. Thus, the selection will have to be newly defined. Whereas, when selecting sub-objects, which are selected by VOL.SELECT, it has the advantage that the selection area will stay intact even when the mesh is changed later.*

**Assigning Waves**

In order to avoid unnecessary calculations, it is worthwhile to limit the wave effect to the area near the camera.

In 3ds max this limitation is done via the modifier VOL.SELECT. By subsequently adding noise modifier, the effect is to the area previously selected.

Settings for Noise:

*FRACTAL* active, *ITERATIONS* 6,0, Z: 30,0 (in the final edition only 10).

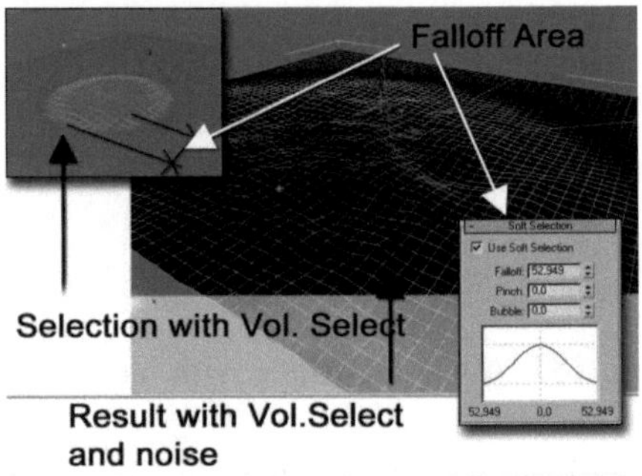


**Fig. 189.** Volume Selection and Noise (*VOL. SELECT – GIZMO VERTEX* and *SELECT BY SPHERE*) This way the noise effect is only assigned to the area selected

[4] Sub-element index – at the moment of generation every element of a mesh or polygon is given a number, the index.

There is one problem when using standard noise (whether fractal or not) - the direction of the waves still needs to be taken care of. This can be done by slightly rotating the Noise-Gizmo, thus tilting its Z-direction (direction of the wave).
By now animating the phase of the noise and tilting the Gizmo in Z-direction, a wonderful waving effect is achieved, with the waves all following one specific direction.
For additionally generating crossing waves, the rotated Noise-Gizmo is simply copied and rotated once more by 180° (or any other value) around the World-Z-axis.
By changing the size of the waves and the phase of the second noise, those small ripples can be animated very easily.

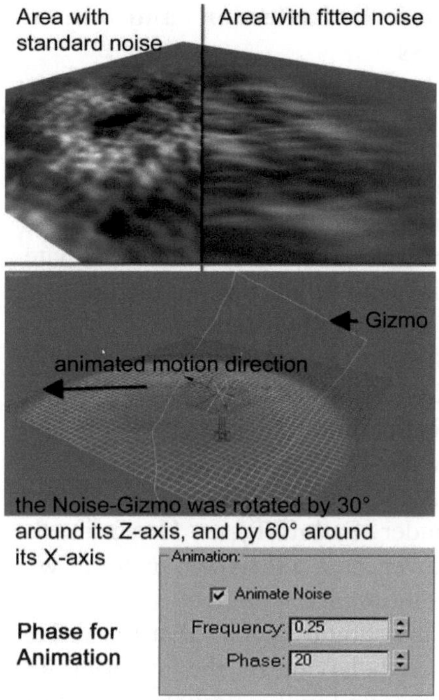


**Fig. 190.** Changing the standard noise by rotating the Gizmo

## Background

As mentioned already in the chapter on atmosphere, there is a very simple way via generating a semi-sphere squashed a little in Z-direction and fitted with a background picture.
The semi-sphere (hemisphere in 3ds max) with the fitted background picture provides the picture data for the reflections on the water surface.

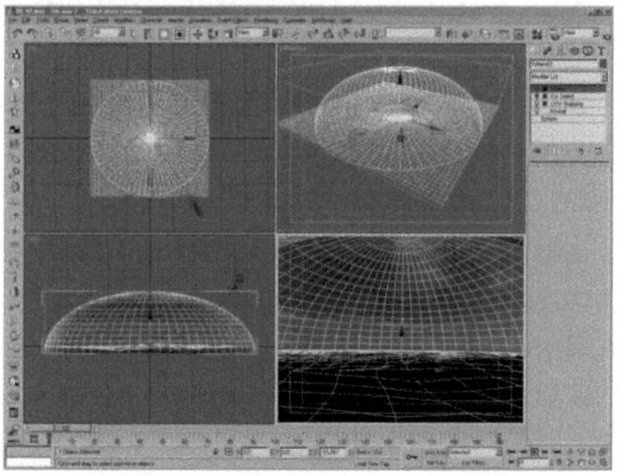

**Fig. 191.** Installing a semi-sphere for the background

**Material/Reflection and Glossiness**

The most important part of a water surface is its material. By installing the right map, you will see that the behavior of reflection and refraction are suitable for the scene.

In the example, the standard material with Blinn Shader includes the following **glossiness:**

*SPECULAR HIGHLIGHTS: SPECULAR LEVEL* 200, *GLOSSINESS* 30 and as **Reflection** 70% and *FALLOFF-MAP* in the *REFLECTION MAP*.

In the Falloff-Map, enter Fresnel under Falloff-Type. Here, in the lower map, a Raytrace material was additionally applied.

This should explain that the reflection behavior of the material just generated will follow the law of the French physicist depending on the distance to the camera and the viewing angle to the water surface.

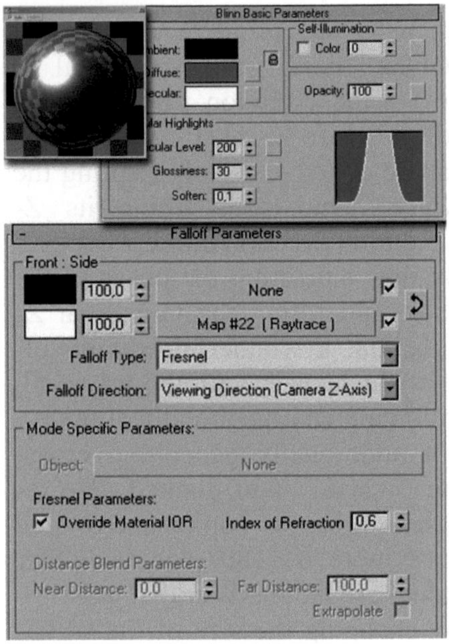


**Fig. 192.** Material-Parameter for Reflection and Glossiness

Except for a small pond or in totally calm weather a natural water surface is very seldom as flat as a mirror. Normally, there is always movement by small or larger waves.

The structure of the surface (the big waves already having been done by geometrical distortion) can be very quickly generated with a Bump Map. In this case too, a Mask-Map is entered into the Bump-Channel - to ensure that the relief does not have to be calculated across the entire surface, and also for obtaining a softer water surface with increasing distance to the camera.

**Material/Relief (Bump mapping)**
Here, a Mask-Map is inserted. This consists of the Bump-Map and the Mask-Map.
The actual relief consists of another map, which has not been mentioned before, the *SMOKE* map.
By reducing the size to a very small value (in the example 0,005), reducing the exponent too (here 0,4), and increasing the number of steps to be calculated (iterations, here 20), very attractive fine-wave/ripple structures can be generated.
When using the Mask-Map, a "mask" is used as an Alpha-Channel. The mask which has been defined as a map will cover the actual map (black – complete cover, white – complete transparency).
By selecting a radial mask (map Gradient Ramp), the bump area fades out centrifugally towards the exterior.

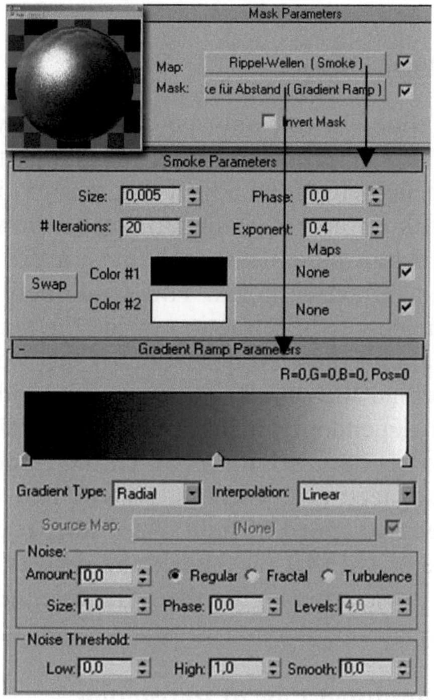


**Fig. 193.** Material-Parameter für Relief und Struktur

By now animating the parameters "phase" of the maps, wave movements even in the material can be simulated quickly and without complications. Although the above example refers mainly to the generation of the surface of the sea or of a large lake, the procedure is nearly the same for smaller bodies of water with a still surface. Here, only the parameters for height and intensity are reduced somewhat. Further variation of the values of the bump maps will quickly result in either calmer, or wilder waters.
Furthermore, the characteristics of the water color and the degree of muddiness still play an important role. Basically, the use of a suitable reflecting material (Raytrace Material or Falloff Map) is the most important prerequisite for generating an authentic looking water surface.
The surroundings and the illumination are also factors which have an influence especially on the objects reflected on the surface. Even though in the example the refraction of light does not play a dominant role, the consideration of the refraction index, as indicated by the refraction characteristics, is an important aid in making visualizations of water convincing.

### *Refraction*

Refraction can only be calculated by a Raytrace procedure. In a refraction, the scene behind the object that is provided with refraction characteristics is shown. All transparent materials have a different refractive index. This value defines the degree of refraction inside the material and thus the distortion of the scene behind the object. The refractive index is dependent on the medium, in other words the medium in which the camera is placed (air, vacuum, water), and the relative speed of light. As a rule, it can be stated that the higher the density of an object the higher the refractive index is.

**Simplified Law of Refraction**

$$\frac{\sin \alpha_v}{\sin \alpha} = n$$

$\sin \alpha_v = \text{Angle of incidence (vacuum)}$

$\sin \alpha = \text{Angle of reflection (medium)}$

The perfect solution is hard to find. Even unit scales of a scene, the characteristics of the environment, and the number of objects in a scene have an influence on the behavior of the water surface.

**Table 6.** Refractive Index

| Material | Index |
|---|---|
| Acetone | 1,360 |
| Alcohol | 1,329 |
| Chromium dioxide | 2,705 |
| Diamond | 2,419 |
| Ice | 1,309 |
| Fluoride | 1,434 |
| Glass | 1,5 - 1,7 |
| Glass (high lead content) | 1,650 |
| Glass (low lead content) | 1,575 |
| Glass (very high lead content) | 1,890 |
| Iodine crystall | 3,340 |
| Carbon dioxid, liquid | 1,200 |
| Crystal | 2,000 |
| Crown glass | 1,520 |
| Copper dioxide | 2,705 |
| Lapis lazuli | 1,610 |
| Air | 1,0003 |
| Sodium chloride | 1,530 |
| Polystyrole | 1,550 |
| Quartz 1 | 1,644 |
| Ruby | 1,770 |
| Sapphire | 1,770 |
| Emerald | 1,570 |
| Topaz | 1,610 |
| Vacuum | 1,0 (precisely) |
| Water | 1,333 |

## Running Water

When generating still water surfaces, the main objective is to design the waves in such a way that the wind is the predominant factor for disturbance. When generating running waters, there are two additional factors that will make the representation a whole lot more complicated. These are the fringe areas between the body of water and the embankment and the running direction of the waves.

Furthermore, rivers, brooks or streamlets and channels are characterized by a considerable greater amount of suspended matter on and just under the surface than for instance the rough surface of the sea.

Not to mention the eddies forming near the banks, and the change in color or flowing direction caused by shallows.

And this is the point: how do you generate a running body of water which looks convincing?

Fig. 194. Running water

Foremost, we are looking for

- The geometry and a suitable shape of waves
- The construction of the materials, and the generation of suspended matter

## Geometry and the Shape of Waves

There are two factors that give running water a special surface. These are firstly the movement of the water due to the existing slope, and secondly the movements on the surface, i.e. waves caused by the wind, as mentioned earlier for still waters.

The movement of the water along the flowing direction (always downhill) is rather constant and even ( an exception being, of course, mountain streams, and suddenly appearing flood waves), whereas the mini-waves appearing on the surface due to the wind can vary considerably.

**Fig. 195.** Example of a scene with running water. In order to emphasize the reflections on the water surface, the trees were added as simple “billboards”.

In the above illustration the flow direction of the water as well as the additional wave formation and the more quiet zones near the embankment were taken into account.

In this scene, the typical visualization problems of a small running water are demonstrated.

Important are for instance the direction of the waves which follow the direction of flow, and the transitional area between the banks and the reflecting water surface. A mistake will soon become obvious when different polygon surfaces are blended.

Another problem area is the embankment area with bays where there is nearly no current and only minimal wave formation. In an example, this could look like this:

**Defining the Surroundings**

A body of water has to be inserted into an existing geometry representing the outline of a landscape.

As in the previous example, a semi-sphere with a sky texture defines the surrounding area. The landscape itself was generated from a simple box via selection of single points (*SOFT SELECTION*) followed by a shift in Z-direction.

The units are defined in meters. The measurements (length/width) of the terrain are 280,0/350,0. Subdivisions are 60 each in X and Y direction. The maximum height of the terrain is 42,0 m.

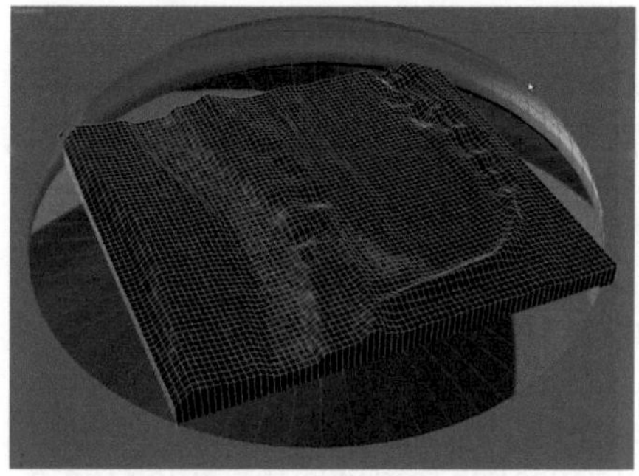

**Fig. 196.** Terrain and sky

**Fig. 197.** Generating the water surface by using a plane

**Generating the Water Surface**

A plane is used as a surface for the running water.

The plane for the water surface should have a sufficiently fine tessellation to ensure the flawless representation of the wave structure later on. The length/width of the plane is 350 x 105. In the present case the *RENDER MULTIPLIER DENSITY* was set at 10. Thus the number of segments is multiplied by the factor 10 at rendering.

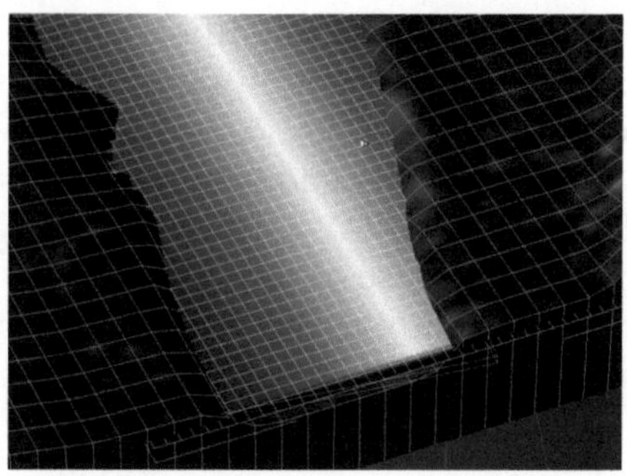

**Fig. 198.** A sufficiently high resolution is important

**Waves in Flow Direction**

A very important aspect is the decreased, slowed down flow in the embankment areas. Thus it makes sense to select only the central areas of the river and to minimize the wave form towards the sides.

A strip in the middle of the river is selected by *VOL.SELECT*, which gets weaker towards the sides via a defined *FALLOFF*.

Depending on the slope and the course of the river it may be more sensible to select the vertex points directly, following the river bends.

Additionally adding a NOISE modifier the effect will be restricted to the previously selected area.

Settings for *NOISE: SCALE* 20,0, *FRACTAL* active, *ITERATIONS*: 6,0, Z: 1,8 (in the example to the right increased to 10 in order to emphasize the effect).

Like in the previous example, the Gizmo of the noise-modifier is now rotated and animated in flow direction.

By adding yet another modifier (*BEND* or *FFD* – freeform deform.) the course of the river can be adapted slightly to the bends and to the slopes.

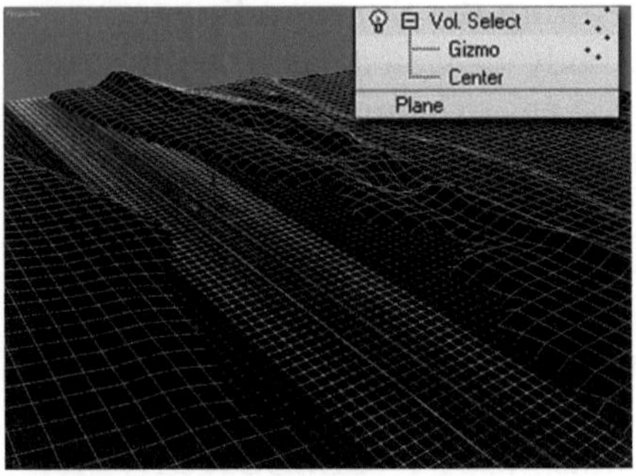


**Fig. 199.** Assigning the volume selection

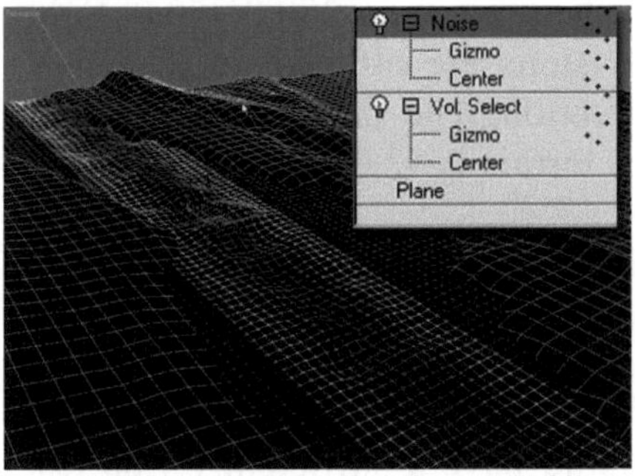


**Fig. 200.** Noise modifier on top of volume selection

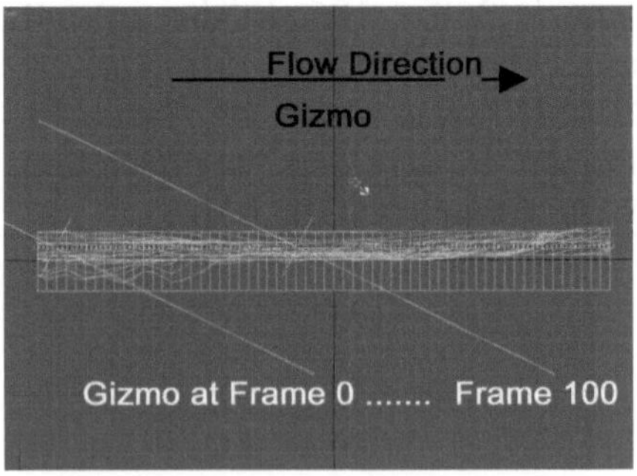


**Fig. 201.** Rotating the noise-Gizmo and animating in flow direction

### Water Surface

After the wave geometry has been assigned, the next important step is the generation of a "suitable" water material.

### Standard Material

An opaque water surface with a correspondingly high reflective behavior can be generated quickly by the following parameters:
*AMBIENT/DIFFUSE RGB: 0,0,0,*
*SPECULAR RGB: 255, 255, 255*
*SPECULAR LEVEL: 200*
*GLOSSINESS: 90.*

### Reflection

For the reflection map, again a *FALLOFF-MAP* with *FRESNEL* as *FALLOFF-TYPE* and *FALLOFF-DIRECTION*: *CAMERA Z-AXIS* is generated. It is important to include a Raytrace-Map in the white slot of the Falloff-Map, to take care of the reflections of the environment/surrounding area on the water surface.

### Relief of the Waves

In order to finally finish the water surface, now only the relief is missing, which will provide the small ripples, if they are to be included. A *MASK-MAP* is added to the *BUMP-MAP*. Just like in still waters, this is composed by a *SMOKE* and a *GRADIENT RAMP-MAP*.

By animating the phase in the *SMOKE-MAP*, the flowing behavior of the ripples can be influenced.

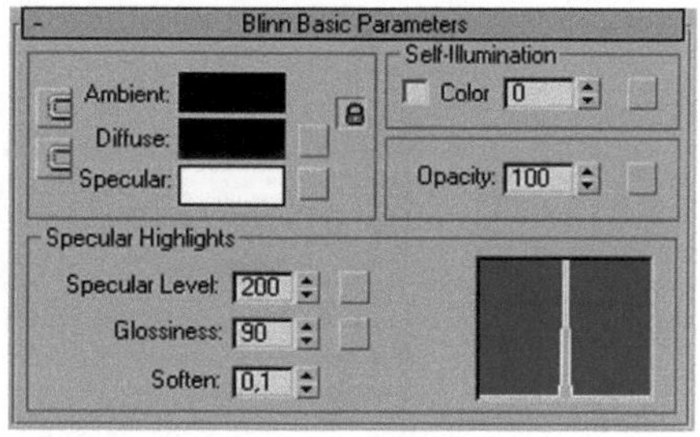


**Fig. 202.** Standard-Material with Blinn Shader

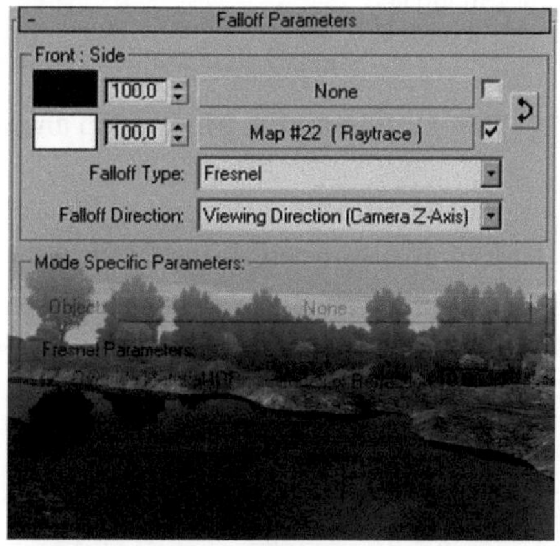


**Fig. 203.** Falloff-settings and reflection

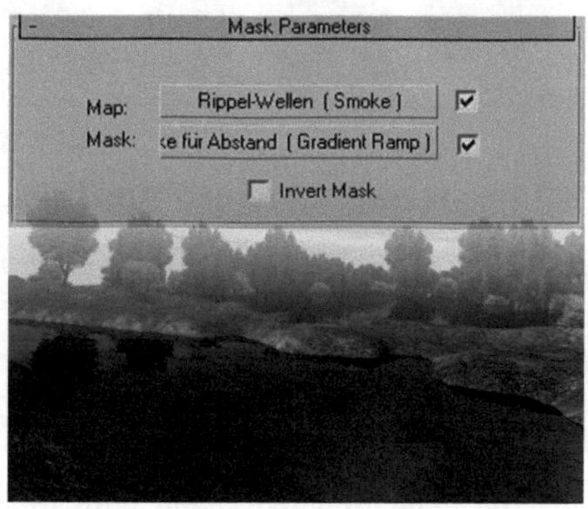


**Fig. 204.** Mask-Map as Relief

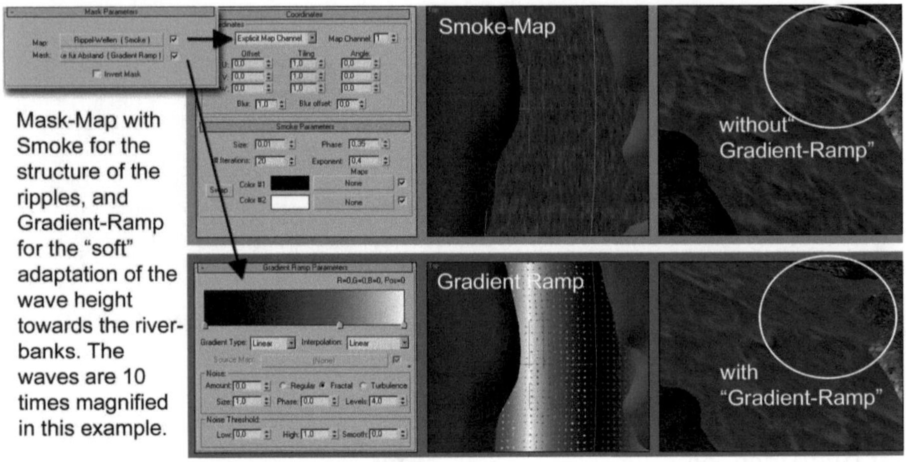


**Fig. 205.** Designing the Mask-Map and adapting the embankment areas

Depending on the purpose of the representation, it may be well worth the effort to provide the edges between the water surface and the riverbanks with some plant growth. This could be grass, shrubs, cat's tails, etc. This has the advantage of hiding foam areas and avoiding sharp edges in an elegant way.

**Fig. 206.** The finished scene with grass and plant growth to cover the line of intersection between water and embankment

As always, such an example can only provide a small glimpse into the actual work, by pointing out the important aspects for the representation of running waters. Anybody studying this subject in more detail, will develop own ideas, recipes and procedures in the course of time. There is no prescription for the "correct" representation of running water. But there are many ideas on how to get there.

## Gushing/Falling Water

Gushing or falling water has a further important aspect which unfortunately cannot be covered by grass or hidden by other tricks: it is the problem of spray and spume.

In running water, the visualization is focused on the representation of the surface, the geometry of the waves, and the reflection. Falling water, however, due to the large percentage of oxygen, is very foamy, full of air bubbles, which give the impression of white particles.

When falling over a sharp edge, as in the picture in the middle, the water stays quite clear. It comes pouring off the edge in one body of water, falling downward in a compact stream, undisturbed. The problem area for visualization, in this case, is located outside the picture segment, namely below left, where the falling water reaches a basin.

The lower picture shows pure spray/spume and water particles enriched with oxygen, bubbles consisting of white parts and of haze.

**Fig. 207.** From top to bottom: Dam wall near Kehl (Germany); Waterfall in the courtyard of the Salk Institute, La Jolla, California, Luis Kahn, 1965; Waterfall on La Gomera (Spain)

The top picture shows a mixed form. In the upper part flowing over the dam wall, the water is still a compact fluid, until, at about a third of the way down, it separates from the wall, filling up with air bubbles and appearing white and foamy.

## Water running over an Edge

The mini-waterfall in the picture of the Salk Institute is to be animated. The exercise is quite simple, because there are distinct geometries, and the form of the water flowing over the edge has already been defined.

### Tipp Matte Material

*Often there is a request for blending a 3D object into a background picture. For achieving this, there is a very effective tool: the Matte-Material MATTE/SHADOW. This material is able to confer various "specific" characteristics to any 3D objects. An object with a matte material can for instance be transparent and still cast a shadow.*

**Fig. 208.** Use of a Matte-Material for blending 3D objects into a background picture.

**Tip Focal Length - a Reminder**

*When you have to fit a camera to a background, it helps to remember that most picture motifs are taken with the standard focal length of 35 -50 mm.*

**Background Picture in 3ds max**

First, a suitable background picture is selected in Max, under *VIEWS • VIEWPORT BACK-GROUND*, then the option *LOCK ZOOM/PAN* is activated, followed by *ASPECT RATIO MATCH BITMAP*. This way, the picture is adapted to the actual view perspective, and automatically adapted, during zooming, to the respective zoom factor. Although the background picture is now visible in the scene, this same picture has to be selected once more as *ENVIRONMENT MAP* under *RENDERING • ENVIRONMENT*, to ensure its appearance during rendering.

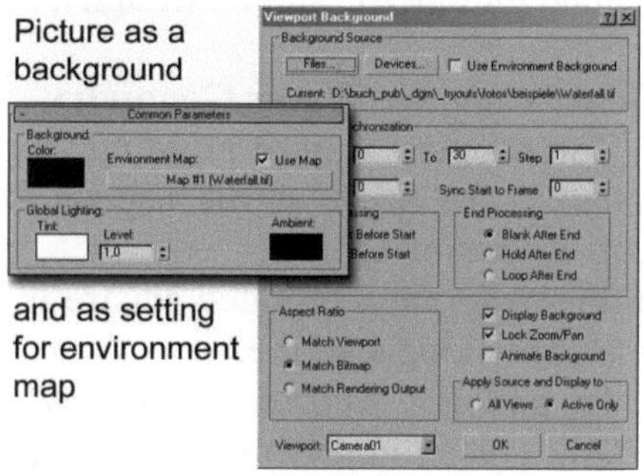


**Fig. 209.** Standard Material

Because on the one hand the water in the scene is partly hidden behind the construction, but on the other hand it is, itself, covering another part of the construction, it is time now to apply the previously demonstrated little ploy for integrating 3D models into background pictures. The ploy is called *MATTE-MATERIAL*.

**Fig. 210.** Simple Geometry

The visible part of the fountain of the background picture is designed by using simple geometries. Then the camera is aligned to that of the background picture, so that the geometry just generated is congruent with the construction on the background picture (Figure 210).

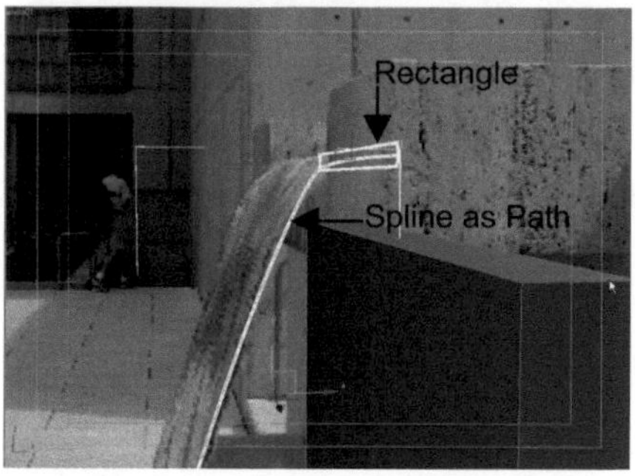


**Fig. 211.** Cross section (shape) and path (spline) to create a loft-object

By now providing this geometry with a *MATTE-MATERIAL* it will cover any further 3D objects within the scene, but still allow the background picture to "shine through".

**Fig. 212.** Smoothing down

### Generating the Water Geometry

The simplest way of building the water in the scene is the use of a Loft-Extrusion Object.
In the example a rounded square is extruded along a spline.

By adding a *NOISE-MODIFIER* to this 3D object, and by animating its *GIZMO* (see Fig. 212) in flow direction, the impression of running water is achieved quickly and convincingly.

Afterwards, in this case too, a further refinement of the noise-animated mesh by a suitable smoothing out (in the example with *TURBO-SMOOTH*) is recommended.

When now assigning the respective material (here the same material was used as in the previous example of the running water), falling water is generated quickly and without too much effort.

**Fig. 213.** All objects blended in

**Fig. 214.** The rendered result

### hint Meta-Balls or Blobmesh

*An alternative for modelling the running water would be the application of so-called Meta-Balls (in 3ds max these are called Blobmesh). Here a particle system is generated first, and then the single particles are joined together to form a continuous fluid object.*

*The advantage lies in the use of a particle system with all its inherent controllable parameters. However, the problem is that Meta-Balls take an incredibly long time to calculate. If feasible, in this case the use of a simple geometry is to be preferred.*

## Waterfall

When trying to simulate a natural waterfall, the use of a simplified geometry like in the case of water falling down over a sharp edge does not lead anywhere.

A waterfall is full of whirls, and the reflection is negligible, because there is no water surface in a piece large enough to make the reflections clearly visible.

In this case, the use of a particle system is a must!

When planning to design a waterfall tumbling down a rocky cliff (like in the example), the following steps can be taken:

- Generating an event-based particle system
- Generating an event which causes a particle to burst into smaller particles after hitting the rock
- Generating an exterior force for the simulation of gravitation
- Generating a deflector which prevents the waterfall from penetrating the cliff
- Generating a suitable material
- Assigning motion blur

One problem arising when following the above method is to make sure that the wall of the cliff remains impenetrable, in other words the use of a deflector to prevent the penetration of water into the rocks. Furthermore, the use of gravitation will require additional calculations, which can turn out to be a real obstacle when the animation has to be finished within a short span of time. Not to mention the event-based particle system.

The above procedure is certainly very effective, but it will take a lot of time and effort to reconcile all the parameters.

Assuming that the waterfall is already in existence, and all that has to be represented is a continuous flow, a simplified method, requiring a lot less effort, could lead to excellent results as well:

Generating a particle system
Generating a path with the help of a spline defining the direction of the waterfall
Generating a suitable material
Assigning motion blur

**Fig. 215.** Waterfall generated by particles

Because the attention of the viewer is captivated by the chaos of the whirls in the water of a waterfall tumbling down a cliff with a lot of spume and noise, gravitation and the deflector can be left out in favor of a spline. Possible penetrations can be ignored, because the viewer cannot really detect them. Thus this simplified method can lead to a very efficient result. By adding a bit of volume fog to the area of the falling water, the completed representation will be even more convincing.

**Environment for the Waterfall**
The geometry in Fig. 217 was generated from a box, by various distortions and adaptations. In the middle of the cliff, there is a groove corresponding to the course of the waterfall.
Following the groove, a spline was generated, serving as a path for the particles to be generated next.

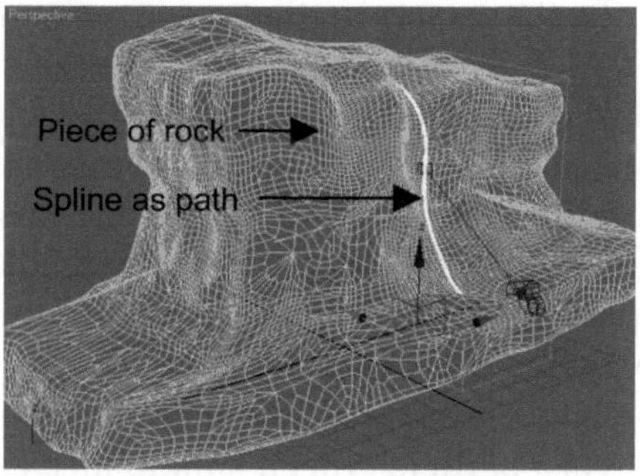


**Fig. 216.** Boulder with a path for the waterfall

**Particle System BLIZZARD**
A simple particle system with the possibility of instancing any geometry as a particle is quite sufficient for the waterfall.

In the example, the system BLIZZARD was selected. This includes an emitter object that is adjusted, in form, size, and angle, to the start and the course (spline) of the waterfall.

In the example, 800 particles are "born", at the speed of 50 units per frame, and end their life after 40 short frames.

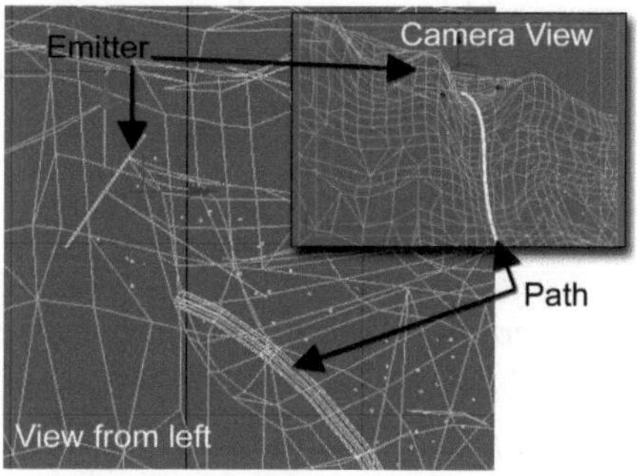


**Fig. 217.** Particle system Blizzard aligned to a spline

Connecting the particles to an outside force enables them to follow the defined course.

The force is called a Space Warp (in 3ds max).

*SPACE WARP • FORCES • PATH FOLLOW*

Space warps correspond to outside forces, like for instance gravitation, wind, etc. which are particularly suitable in connection with particle systems.

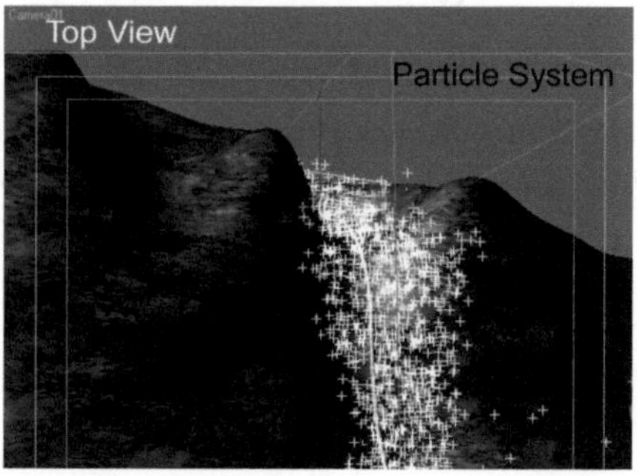


**Fig. 218.** Shaded representation of the particle system

The use of such a space warp enables the particles to follow any chosen path.

In the example, the space warp *PATH FOLLOW* was activated under *SPACE WARPS • FORCES*, to which the previously generated spline was then assigned. By linking the particle system to *PATH FOLLOW*, the generated particles will follow the path according to certain criteria. These criteria can define, for instance, the possible action radius of the particles, the animation length (i.e. how long the particles will follow the path), the speed distribution and the rotation around the path selected. Although the particles will penetrate the rock, this "mistake" is negligible and can hardly be noticed due to the turbulent motion of the water tumbling down.

As a particle type, a simple face can safely be selected. A plane made up of a simple polygon needs considerably fewer resources than, for instance, a sphere as a particle type.

In this context, another very important aspect is *MOTION BLUR*. The faster an object moves in front of our eyes, the blurrier it appears. The particles are assigned a *MOTION BLUR* as characteristics via the right hand mouse key *PROPERTIES*:

### hint Motion Blur

*A fast moving object can easily be generated via the so-called object motion blur. The motion blur is generated when rendering. Several copies of the object are calculated between the frames, and then the entire scene is rendered once more. The object motion blur is not influenced by the motion of the camera. As opposed to this the picture motion blur will simulate the lack of focus similar to a blur in a picture editing program.*

*The advantage of the picture motion blur is without a doubt the speed of rendering, because a lot more calculation effort is required when determining the motion blur for each single object.*

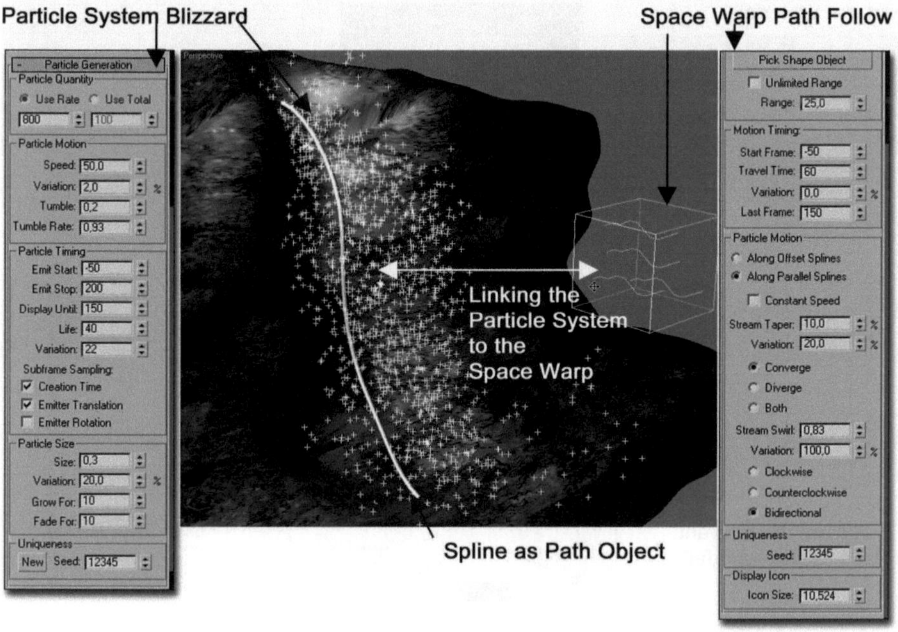


**Fig. 219.** Interdependency of Particle System and Space Warp

Another important aspect is of course the material to be used.
The conditions to be fulfilled could be formulated more or less like this:

Spume is white
Blur is important
Haze and fog play an important role
Tumbling water looks more chaotic than symmetrical
In a *spumy* waterfall the reflection of the water plays a subordinate role

A “genuine mirror-effect” such as when using a Raytrace-material may not be necessary, however, the water material should have a high gloss coloring. This implies that the value of a standard material (Blinn) should have a very high setting for *SPECULAR HIGHLIGHTS*.
The remaining Map-Channels that are important for the material can be limited to the *DIFFUSE*, *OPACITY*, and *BUMP* Channels.

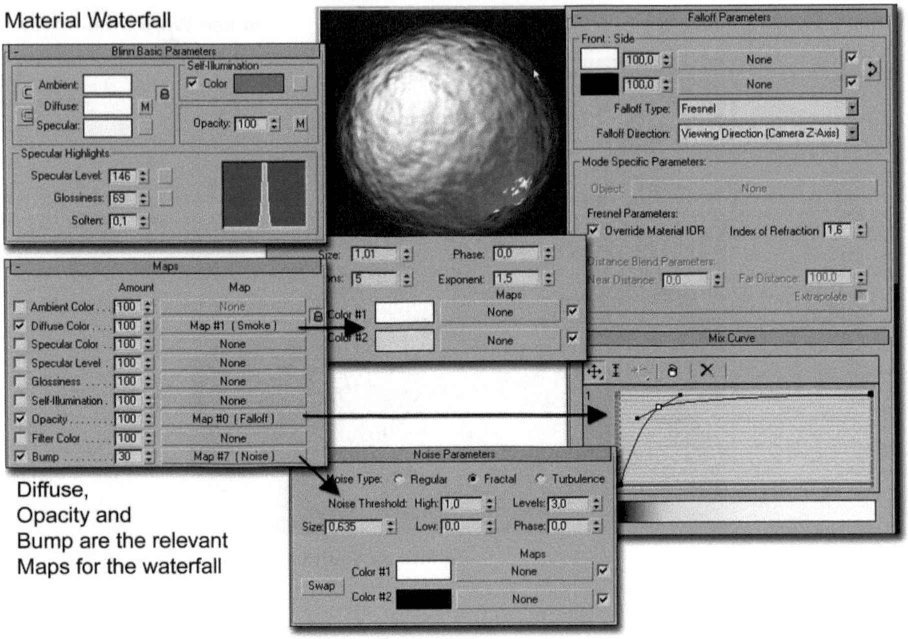


**Fig. 220.** Overview of material parameters

Water has the characteristic transporting light as it were just beneath the surface. Depending on the incidence of light, this can lead to fluorescent effects in the spumous areas. Because the waterfall consists mainly of spume, depending on the appearance of the entire scene, a self-illumination of the water material would be an effective complement. A little additional atmospheric fog (volume fog, the generation of which was explained in the previous chapter) will make the waterfall even more convincing.

It should be obvious that the "proper" generation of a waterfall, as near to reality as possible, would take a lot more time than the example above. Unfortunately, a detailed introduction into event-based particle systems would exceed the scope and size of this book.

## Tip Speeding up the Rendering Process

*For speeding up the rendering process it may be helpful to de-activate the setting "receive shadow" (right click selected object > properties). The option of casting a shadow for an object should in any case be tested beforehand, depending on the particular application.*

## Border and Transition Areas

The last variation of tumbling water will deal with the transition between running and spumy water, as can be seen in the example of Fig. 207, the dam wall. In a slightly modified example of running water a solution could look more or less like this:

**Fig. 221.** Streaming and gushing outlet with transition area

The scene in this example is the same as in the first example. Only the two borders of the water and of the material were changed. Additionally, the previously generated particle system has been integrated, with slightly modified parameters. Thus the material now used includes the previously generated material for running water, as well as the spume material without opacity, plus a map for the *GRADIENT RAMP* defining the transition of the two materials.

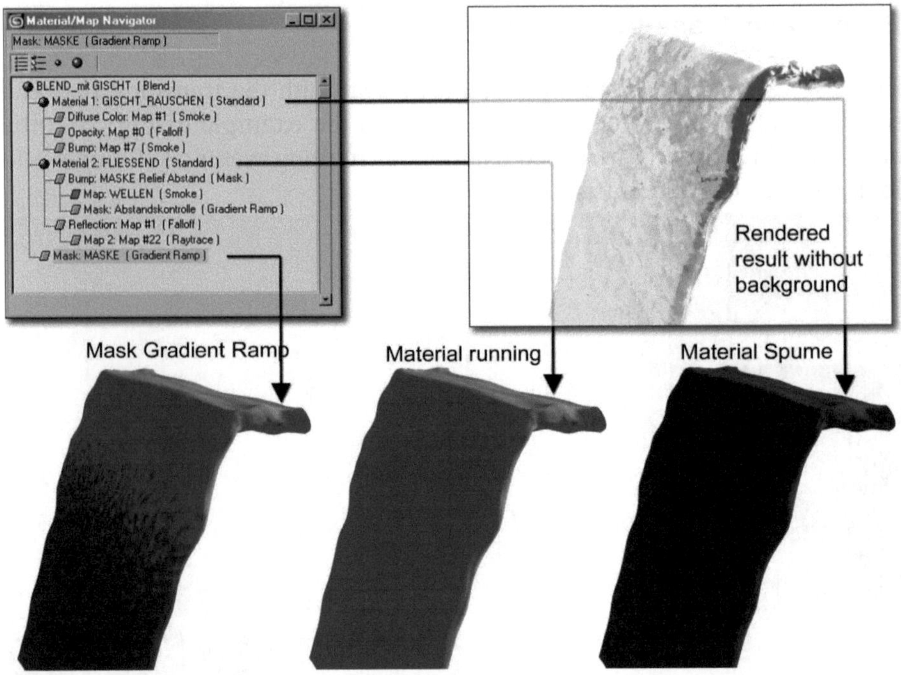


**Fig. 222.** The Blend-Material for flowing and turbulent outlet with corresponding mask

**Fig. 223.** The running water with transition area and additional particle system

The particle system was positioned in the upper third of the geometry of the running water and aligned to a spline like in the previous example. In this case the extrusion path of the running water serves as direction indication pointing the way for the particles.

## Light Reflections by Caustic Effects

Caustics are those effects that are caused when rays of light hit a water surface and are reflected, and these reflected rays illuminate the surrounding area. Caustic effects are also caused by refraction. In the picture to the right showing a drop of water you can see the shadow caused by the source of light, but you can also see the spot of light which appears right in the middle of the shadow and which has been caused by the refraction of the drop of water. This phenomenon is called caustic effect.
Not long ago, the calculations for caustic effects belonged to the domain of very expensive rendering engines. Today, the possibility of simulating these effects has become a standard feature of nearly every 3D program.
When rendering the scene without any consideration of caustics or global illumination, the result will look more or less like this:

**Fig. 224.** Grotto with water without reflections and caustic effects

**Fig. 225.** Grotto with water rendered without reflections and caustic effects

The scene in the above example shows a grotto with a water surface. The water surface is illuminated directly and from above by one single source of light (a spotlight). Similarly to the previously described water surfaces, a noise-map for light wave formation, and a Raytrace-map for reflection have been assigned to this water surface.

A very simple way of simulating the effect caused by caustics is the use of an additional light source to serve as a projector for generating caustic reflections.

### hint Physically correct or Fake?

*Caustic calculations, similarly to calculations of global illumination, require time and additional computing. As always, the type of simulation depends on the requirements for the design of the scene. If it is a case of having to show a water surface with caustics during a camera "walkthrough", one can safely use the cheat package with an additional light source. However, if the scene with the water surface generating caustic needs to be shown for a longer time and with a fixed camera, then the bigger calculation effort of using correct caustic effects can definitely be advised.*

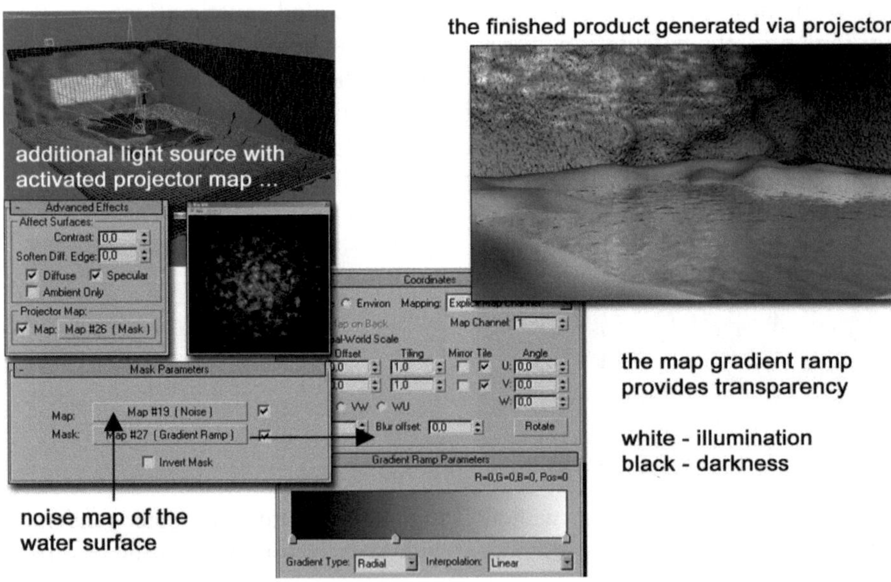


**Fig. 226.** Grotto with additional light source and application of a projector map

In the example, a bump map including a noise map was generated for the water surface. Generating a new mask map and instancing here the same noise map as for the relief of the water surface, assigning a radial gradient ramp map as mask and then using the entire mask map as projection for the additional light source illuminating the area where the caustic effects will be expected - this will certainly be sufficient for a quick representation. The advantage of the instanced noise map is the fact that, in an animation, the projection on the wall generated by the additional light source will behave exactly like the moving water surface.

Admittedly, the physically simulated effect will, in most cases, look quite different, and the optimal solution will be the use of a suitable program for the correct calculation of caustics.

The following picture was generated in 3ds max7 with the renderer Mental Ray[6]. One single light source, provided with a sufficient number of photons, has achieved convincing and "near-correct" results of the effect in an acceptable rendering time.

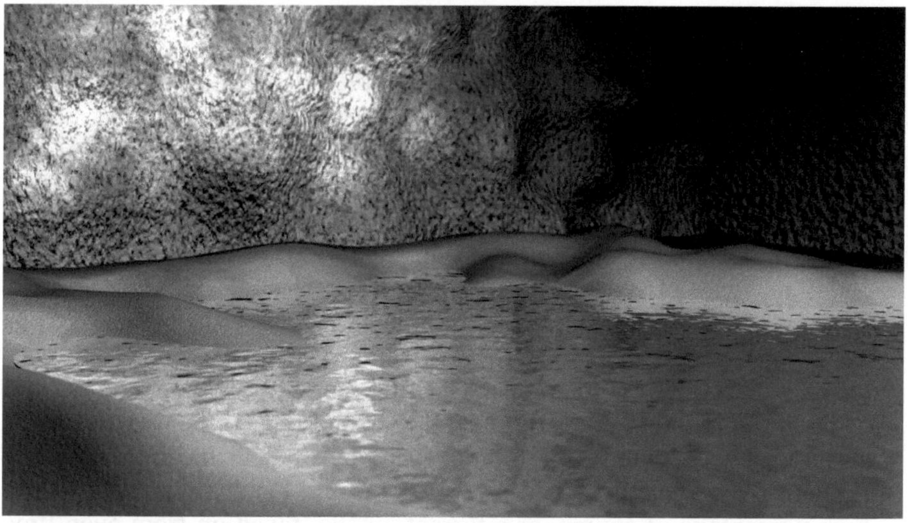

**Fig. 227.** Grotto with Mental Ray and physically correct calculation of the resulting caustic effect

[6] http://www.mentalimages.com

## Summary

Water is an incredibly complex element. In this chapter only a short overview could be given of some aspects relevant for the landscape architect.
In order to study the fundamentals more thoroughly, it is recommended, as in nearly all areas of visualizing natural phenomena, to take a very close look at nature and try to analyze very thoroughly the significant characteristics of the medium to be represented, to be able to achieve a convincing 3D visualization. The Fresnel Effect, for instance, ensures the correct type of reflection of open water surfaces. Reflection and refraction of water can be simulated convincingly by Raytrace materials.
Some of the most important topics for the representation and simulation of water and methods of visualization are:

The representation of open water surfaces, like for instance the surface of the ocean or of a big lake. The various types of waves are important here.

The representation of running water for rivers, streams and brooks, and the "correct" use of various materials and noise distortions for simulating the flow effect. The transition areas near the river banks are often problem areas, but can be covered up by the skilful use of grass, reeds, and other types of vegetation.

Tumbling and falling water which can be observed in dam constructions and in the design of man made waterfalls, can be visualized and animated roughly by two methods. The first method is the generation of a geometry simulating the water, and the second type is the use of particle systems. The geometry of running water is used in the case of water flowing over a sharp edge. For representing a natural waterfall with a high percentage of spume, or a torrent, or the transition from even flow to a turbulent flowing behavior, however, the use of various types of particle systems is necessary.

The combination of water and light cause so-called caustic effects. These effects can either be generated manually by using suitably manipulated light sources, or by physically correct simulations using the corresponding rendering engines. As in the case of light sources, the choice of the right method will always depend on the time available and on the type of scene that is needed.

# Rendering & Post Processing

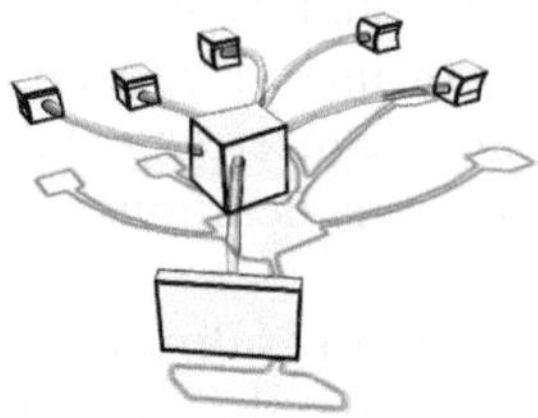

*Where is the „render cool picture" button?*

## Rendering?

Rendering is a word used in various differing contexts. It implies picture calculation or generation, and is used as a term in all those programs, which generate, by way of complex calculations, new pictures, movies or sounds from digital "raw material" of any kind.

The term is used in computer graphics as well as in classic layout. Here, however, it is used for coloring of sketches or scribbles.

Insiders talk of rendering in the field of composing a picture for a website via a browser as well as for calculating a sound sequence for an audio-program. In short, the term rendering is used often and is very complex.

With respect to 3D visualizations, rendering is the special procedure, which generates a finished picture, a sequence of pictures or a movie, out of the designed 3D scene, the illumination and all the materials.

The question is now, what needs to be observed during rendering and subsequent (not always necessary) postproduction?

Postproduction means film editing, color correction, cutting, and/or later optimization of picture and movie material. Most 3D programs offer quite a few postproduction tools, and if this is not sufficient for you, you may want to turn to specialists in this field, like for instance Combustion or After Effects.

The main points in this chapter can be summed up as follows:

- Some information about picture- and movie-formats. Which types of pictures do exist, and what are Codecs, etc.? Is it better to do the calculation for the movie sequence or for the single picture?
- What are rendering types
- Image control
- Optimizing rendering procedures
- Rendering effects and environment
- Working with layers
- What are the possibilities for postproduction of pictures in a picture-editing program like Photoshop?

The quality of a generated 2D image depends on various factors. These are not factors concerning the model or the examples mentioned here, but factors concerning the quality of the picture which can be described by three considerations:

**Image resolution**: How many pixels will you grant your picture, or, in other words: what are the general conditions for the general setup of the 3D animation or the respective still frame?
**Type of data file**: Which type of format is to be used? Should it be a compressed picture format entailing a certain loss, or would it be better to use a loss-free format?
**Type of presentation**: Will the finished movie sequence be shown in a cinema, will it be a home video, a multi-media CD, a movie (strip) for streaming contents in the web, or are the pictures produced for an edition in the printed media or for a poster?

Within a project, each one of these considerations represents a constraint for the general conditions. For private use you have all the possibilities at your disposal, but in the end it pays to keep up your discipline, because more often than you expect, your professional input will be based on work produced at home, on rainy weekends.

## Some General Remarks: Pictures and Movies

Still frame or movie? There is no contradiction here, because a movie is nothing other than a series of single frames. These are mostly saved in digital video formats like AVI or MOV.

AVI (Microsoft video format) or MOV (Quicktime movie format by Apple) are both so-called container formats. These container formats can be compressed by different methods. This type of video compression is called CODEC (enCOder/DECoder).

It is good to start with still frames. There are different formats in this field. Every picture format has its own history and its preferred area of use. Deciding which format is to be taken for which purpose depends on what it will be used for. In the following, there is an overview of the most current formats plus an indication of their predominant area of use.

## Image Types and Formats

**Table 7.** Table Image Types and Formats

| Name | Term/Definition | Compression |
|---|---|---|
| JPEG, JPG | Joint Photographic Experts Group | Lossy Compression (JPEG) |
| JPEG is short for Joint Photographic Experts Group and is a picture format with lossy compression. JPEG is the common format for the representation of photographs and other semi-tone pictures in HTML[1] files in the World Wide Web and other online services. The JPEG format supports CMYK, RGB and gray scale color modes. It does not support Alpha Channels. In contrast to GIF, JPEG contains the entire color information of a RGB picture. JPEG uses a compression for reducing the file size by recognizing and deleting data which is not relevant for the picture. When you open a JPEG picture, it is decompressed automatically. The higher the compression, the lower the quality of the picture; the lower the compression, the higher is the quality of the picture. In most cases a compression with the highest possible degree of quality will produce a picture that is almost identical to the original. | | |
| GIF | Graphics Interchange Format | Lossless Compression (LZW) |
| GIF is short for Graphics Interchange Format, a picture format with loss-free compression (LZW). It is the most popular format for the representation of indexed color pictures[2] in HTML files in the World Wide Web and other online services. GIF is a LZW- compressed format which was developed to reduce the file size and the transmission time via telephone lines as much as possible. The GIF format does not support Alpha Channels. | | |

[1] HTML- Hyper Text Markup Language: HTML is used for describing the logical components of a document.

[2] Indexed colors are limited to a maximum of 256 colors.

| | | |
|---|---|---|
| PNG | Portable Network Graphics | Lossless compression |
| PNG is short for Portable Network Graphics and is a picture format with lossless compression. It represents an alternative to the JPEG-format. Just like GIF, PNG is used for the lossless compression and representation of pictures in the World Wide Web and other online services. As opposed to GIF, PNG supports 24 bit pictures and generates transparent background areas without sharp edges. The PNG format supports gray scale and RGB files with an Alpha Channel, and indexed color and Bitmap files without Alpha Channel. | | |
| TGA (Targa®) | Truevision Format | Uncompressed/Lossless Compression (RLE) |
| TGA is short for Truevision Advanced Raster Graphics Array and is a picture format available compressed with lossless compression (RLE), as well as uncompressed. TGA was developed for use with systems working with Truevision video cards; it is often used in MS-DOS color programs. The Targa format supports 32 bit RGB files with a single Alpha Channel, and indexed color, gray scale, and 16 Bit and 24 Bit RGB files without Alpha Channel. TGA formats are often used for rendering of still frames. | | |
| RPF | Enlarged SGI Format | Uncompressed |
| RPF is short for Rich Pixel Format and is an enlarged uncompressed SGI format. While still under the name of Kinetix, the suppliers of 3ds max took a closer look at the RLA format and decided to upgrade it. Thus, additional channels, especially for motion data (motion blur), were built into the existing format and integrated into Max as a Kinetix specific RLA format. The development went even further, and the RLA format was offered on the market as RPF. Just like the RLA format, RPF is a format which was developed specially for the postproduction of 3D sequences. Additional upgrades are, among others: Transparency, Speed, Sub-Pixel Weight, and Sub-Pixel Mask. | | |

| RLA | SGI-Format | Uncompressed |
|---|---|---|

RLA is short for Run Length Encoded Version A and is a widely used uncompressed SGI format. It was originally developed by Wavefront. The RLA format supports 16 bit RGB files with a single Alpha Channel. RLA is an excellent format for the post production of 3D visualizations, because basic information layers can be saved in this format. Apart from this, RLA offers seven additional channels:

1. Z Depth: indicates Z-Buffer information repeating gradients from white to black. The gradients indicate relative depth of the object in the scene.

2. Material Effects: saves the effect channel which is used by materials assigned to particular objects in the scene.

3. Object: indicates the object channel ID of the G-Buffer. The G-buffer ID is assigned using the object properties dialog.

4. UV Coordinates: indicates the range of the UV mapping coordinates as color gradient.

5. Normal: indicates the alignment of the normal vectors as a gray scale gradient.

6. Non-Clamped Colors: indicates areas in the picture where colors exceeded the valid color range and were corrected.

7. Coverage: this saves the coverage of the surface fragment from which other G buffer values (Z Depth, Normal, and so on) are obtained.

| TIF, TIFF | Tagged Image File Format | Compressed or Uncompressed (LZW) |
|---|---|---|

TIF,TIFF is short for “Tagged Image File Format” and is certainly one of the most widely used picture formats, in all areas. It is available compressed (mostly LZW) as well as uncompressed. TIF is used for the exchange of files between various programs and platforms. TIF is a flexible Bitmap format which is supported by practically every painting, picture editing and page layout program. Nearly all Desktop Scanners can produce TIF pictures as well.

The TIF format supports CMYK, RGB, and gray scale files with Alpha Channel, and LAB indexed color, and Bitmap files without Alpha Channel. TIF supports LZW compression.

<table>
<tr><td>BMP</td><td>Bitmap</td><td>Compressed or Uncompressed (RLE)</td></tr>
<tr><td colspan="3">BMP is short for Bitmap and is available compressed (RLE) as well as uncompressed. BMP is the standard Windows-Bitmap format and is used on DOS- and Window-compatible computers. When saving a file in this format you can do so either as Format Microsoft Windows, or as OS/2, and with a color depth of 1 bit to 24 bits. For 4 bit and 8 bit pictures you can also select the compression Run-Length-Encoding (RLE); this type of compression is loss-free. BMP is a very suitable format for editing on Windows platforms. The use of BMP is not recommended for data exchange.</td></tr>
<tr><td>PCX</td><td></td><td>Compressed (RLE)</td></tr>
<tr><td colspan="3">PCX is short for Picture Exchange and can be compressed losslessly by the RLE procedure. PCX was originally developed by the company Zsoft as a picture format for the painting program Paintbrush. Most PC programs support PCX. Version 3 does not yet support own color tables/charts. The PCX format supports RGB, indexed color, gray scale, and Bitmap files. It does not support Alpha Channels. PCX supports the RLE compression. Pictures can have a color depth of 1, 4, 8, or 24 bit.</td></tr>
<tr><td>PSD</td><td>PhotoShop File format</td><td>Compressed</td></tr>
<tr><td colspan="3">PSD is the upgrade of file names for graphic files of Adobe Photoshop. It is based on Postscript and is now an established “standard” of picture processing. As a file format for picture editing, the PSD format is becoming more and more popular. The format supports several picture layers, which superimposed, form the actual picture. Each layer can have as many channels as needed (e g R, G, B, Mask, etc.). By providing these various layers, this file format opens a wide field of creative design possibilities, leaving space for numerous special effects. PSD supports all color depths.</td></tr>
<tr><td>PS</td><td>Postscript</td><td>Uncompresed</td></tr>
<tr><td colspan="3">PS is short for Postscript, it is uncompressed and actually a printer’s language; it is a media and resolution - independent format of ADOBE. It was developed for the graphics industry and is still used in the field of printing and the media in general. Postscript describes contents and layout of a page containing text and graphics. Postscript can include vector and pixel information. In this context the format EPS (Encapsulated Postscript) should be mentioned:</td></tr>
</table>

| EPS | Encapsulated Postscript | Uncompresed |
|---|---|---|
| EPS is an upgraded version of Postscript, which offers, apart from the possibility of processing vector and pixel information, also the option of using masks. EPS supports Lab, CMYK, RGB, indexed color, Duplex, gray scale and Bitmap files. It does not support Alpha Channels. | | |

**Table 8.** Procedures for compression of digital images

| Name | Term / Definition | Remarks |
|---|---|---|
| JPEG, JPG | Joint Photographic Experts Group | Lossy compression (JPEG). JPEG compression offers very good results for photographs and colourful pictures. |
| LZW | Lemple-Zif-Welch | Lossless compression. Very suitable for the compression of pictures with large homogenous color areas. |
| RLE | Run Length Encoding | Lossless compression. |

## Which Format is used when?

Naturally, the question will arise of which picture format has to be used for which purpose. Depending on the later use, it is recommended to produce the picture formats to be generated as original data in as high a quality as possible. This will automatically disqualify, so to speak, all picture formats with lossy compression.

It is no problem to reduce a high quality picture format later, but if too many allowances have been made for the memory requirements of the hard drive, the way back will always imply a loss of quality.

It is always advisable to generate a high quality starting size, and to take this as a basis for further use.

**Basic Format**: Good starting formats are TGA or TIF. Both can administer Alpha Channels, and can be compressed losslessly. These starting formats are also suitable for the generation of still frames. The choice between TGA and TIF is a matter of personal preference. The possibility of administering CMYK certainly indicates a close connection of TIF to the print media, while TGA has become an established standard of video editing and postproduction. Because it does not make much

sense to generate CMYK directly from a 3D program (in any case, normally it cannot be done anyway), it is certainly advantageous to confine oneself to the RGB color range. If necessary, it is better to do later processing with respect to color adaptation achieved via a picture software such as Photoshop.
**Additional formats**: Depending on the area of use, the generation of JPEG or PNG is recommended for multimedia applications. Both can be generated from the starting size.

**Tip EPS Format for Printing Sizes with 3ds max**

*If you want to render a picture with particular measurements, choose the EPS format. Here you have the option of a direct input of the lateral measurements as well as the resolution (dpi).*

However, RLA and RPF files normally only make sense when using software like Combustion or After Effects for postproduction.

### Video Formats

Two video formats have already been mentioned: AVI and MOV. These two formats can either be rendered directly from nearly all 3D programs, or via video postproduction. Both formats are so-called container formats. This implies that although on the outside always labeled the same way (*.AVI or *.MOV), they can be compressed by different methods. The compression methods for video formats are called Codec (COmpression / DECompression).

#### *AVI*

Audio-Video Interleaved is the Windows standard format for movies. AVI movie sequences are used in the following areas:

Animated materials in the Material Editor
Animated Backgrounds
Compositing in video post production

The most current Codecs[3] are:

**Cinepak Codec by Radius:** This is the standard codec of most windows platforms. The quality is moderate, but you can be sure that an AVI compressed by Cinepak will run on every computer. Cinepak supports color depths of 24 bit.
**Intel Indeo Video R3.2, 4.5 und 5.10:** A codec with high compression rates and good quality. It is not normally included as a standard, but has to be installed additionally. Intel Indeo supports color depths of 24 bit.
**Microsoft Video 1:** This codec offers a lossless compression and supports color depths of 8 to 16 bit.

**MPEG:** MPEG is short for Moving Picture Experts Group; this is a group of specialists and experts working in the field of standardization of video compression techniques and methods (more information under http://www.chiariglione.org/mpeg).
**MPEG-1:** The picture format is very similar to that of JPEG. This codec has been known for its widespread use for VCD (Video CD) - more popular however in Asian countries than in Western industrial nations.
**MPEG-2:** MPEG-standard for video and audio compression. MPEG-2 is the successor of MPEG-1. The reduction of data volume is achieved not only by compression but also by data reduction. The successor of MPEG-2 is:
**MPEG-4:** MPEG-4 is an internationally standardized procedure for the memory-optimized digital recording of moving pictures combined with a multi-channel sound. MPEG-4 offers a high compression with high quality. Due to the fact that MPEG-4 does not compress every single picture (instead, the compression is achieved as a result of changes from one picture to the next), this codec is not suitable for video cutting.
**DivX:** A freely available codec, which is strongly influenced by the MPEG-4 codec. This codec is characterized by a very high compression and a good quality. DivX is just as unsuitable for video cutting as MPEG-4.
**WMV (Windows Media Video)**: Microsoft's family of video codec designs including WMV 7, WMV 8, and WMV 9. It can do anything from low-resolution video for dial up Internet users to HDTV. Files can be copied to CD and DVD or output to any number of devices. It is also useful for Media Centre PCs. WMV can be viewed as a version of the MPEG-4 codec design. The latest generation of WMV is now in the process of being standardized in SMPTE as the draft VC-1 standard.

[3] Find out more at: http://www.wikipedia.org

**RealVideo:** Developed by RealNetworks. A popular codec technology a few years ago, now fading in importance for a variety of reasons.
**H.263:** Used primarily for videoconferencing, video telephony, and Internet video. H.263 represented a significant step forward in standardized compression capability for progressive scan video. Especially at low bit rates, it could provide a substantial improvement in the bit rate needed to reach a given level of fidelity.

### *MOV*

Apples Quicktime Format was developed at the start of 1997, and is available for nearly all platforms. For viewing the MOV file format, the Quicktime Movie-Player is required. Quicktime can represent and administer more than 70 formats (Codecs) from various fields of use, and includes most conventional multimedia and compression standards.

Furthermore, Quicktime offers the possibility of representing an interactive data format, Quicktime VR. This format enables, for instance, panorama representations of 3D scenes and can be exported from 3ds max.

### *A Movie from Stills?*

The best method for generating animations and their further editing is the generation of sequences of single still frames. By rendering the animation in TGA files, including an Alpha Channel, you will have a very solid basis for further processing at your disposal.

Convert the single frame sequences into the corresponding video formats only when requested.

As a rule, for such conversions a video editing program such as Adobe Premiere, Main Actor, Speed Razor, Final Cut, etc. is used. Most conversions into current formats can also be done directly in many 3D programs during video post processing.

## Image Sizes

Different fields of application require different resolutions. Naturally, when using a chart in a power point presentation, a different resolution will be needed than for use in the print media.

The graphics shown in this book were all generated as gray scale files in the TIF format, with a resolution of 600 dpi, a horizontal dimension of 11,7 cm maximum, and a horizontal resolution of 2764 pixels.

## Still

For still frames there are almost no limitations. The render output size is limited only by the constraints of the respective software[4].

Single images for the print media should be done with at least 300 dpi (dots per inch, or pixels per inch; one inch is equal to 2,54 cm) for color pictures, and at least 600 dpi for gray scale pictures.

*Resolution for representation on a monitor*

$$L_{\text{pixel [required length in pixels]}} = X \text{ [desired length in cm]} \; \frac{96\,\frac{Pixel}{inch}}{2{,}54\,\frac{cm}{inch}}$$

*The resolution of 96 dpi corresponds to the current resolution of PC monitors.*
*Example for a print edition:*
*The desired length of a picture is 24 cm the required resolution is 300 dpi.*

$$L_{pixel} = 24\text{ cm} \; \frac{300\,\frac{Pixel}{inch}}{2{,}54\,\frac{cm}{inch}} \qquad L_{pixel} = 2834\text{ Pixel}$$

*The required setting at rendering is 2834 pixels.*

**Fig. 228.** Determining the resolution in pixels

Since the question of the correct resolution always comes up again, the above formula will be very helpful.

## Movie Formats

For the edition in movie formats it is best to follow the standards which are by now included as standard features in most 3D programs. Most current movie and video formats will already turn up in the rollout list of the rendering dialog, e.g. PAL (Video) with a resolution of 768 x 576 pixels (or DV-PAL 720 x 576), etc.

### hint 25 fps in 3ds max

*When rendering in PAL resolution, keep in mind the corresponding setting for the time configuration of 25 frames per second.*

[4] In 3ds max rendering comes to a dead stop at 8.800 pixels

*The standard setting of 3ds max has been fitted to the American standard of NTSC and 30 frames per second.*

## Rendering Procedure

Although the examples in this book are mainly based on 3ds max, the relevant settings and variables are similar in most 3D programs.

Nearly all settings for the rendering edition can be found in the dialog field menu *RENDER SCENE (F 10)*.

General parameters like single frame, or movie sequence, editing size, and further options like file names, render elements, selection and settings of the actual renderer, are defined here.

### Output Size

As mentioned before, the output size will depend on what you need to be able to do with the data later on.

One value which keeps coming up again and again is the value 640 x 480 pixel. This numerical value originates from the old days of the VGA resolution and can be played on nearly every PC.

### Image Aspect

When specifying the size of an image for the render output, a specific image aspect (pixel width to pixel height), as well as a pixel aspect, is always assigned. The pixel aspect defines the form of the pixels (1 corresponds to a square). These ratios originate from the different video formats.

#### hint Pixel Aspect

*Measurements are not only relevant in the side ratio of the image (frame). Some video formats have identical side ratios, but a different side ratio for the pixels of which the frame is composed. The standard format D-1 (CCIR-601), for instance, produces the same side ratio of 4:3 as the standard formats of Windows, Macintosh, and NTSC, but uses rectangular pixels with a resolution of 720 x 486.*

**Fig. 229.** The picture on the left, with a resolution of 768 x 576 pixels, demonstrates the result of a rectangular form of pixels. The pixel aspect is at a very exaggerated 0,6. The picture on the right has a pixel aspect of 1,0 which indicates square pixels.

## Video Color Check

Depending on the further use of the video production, some colors are outside the valid range of the video format. They tend to blur or fuzz when transferred from computer to video. Pixels containing these "illegal" or "hot" colors are flagged on the sample object. With the option video color check the values of the picture to be rendered can be examined, and either be corrected or represented in black.

## Atmosphere

Atmospheric effects are controlled in the render dialog. Parameters are manipulated and effects are switched on or off.
In test renderings a lot of time can be saved by a short-term deactivation of atmospheric effects.

## Super Black

Super Black is an interesting variation, to be used instead of an Alpha Channel. You want to render a finished movie sequence, and you have a black background in your scene (e.g. object visualizations). By activating Super Black, the background is assigned the color black (RGB 0,0,0), while all other objects in the scene which should actually also be black are rendered in a slightly “lighter” black.

In this way, it will be possible to mask the objects later, i.e. to select ("key") the background for a masking function.

## Render to Sequence

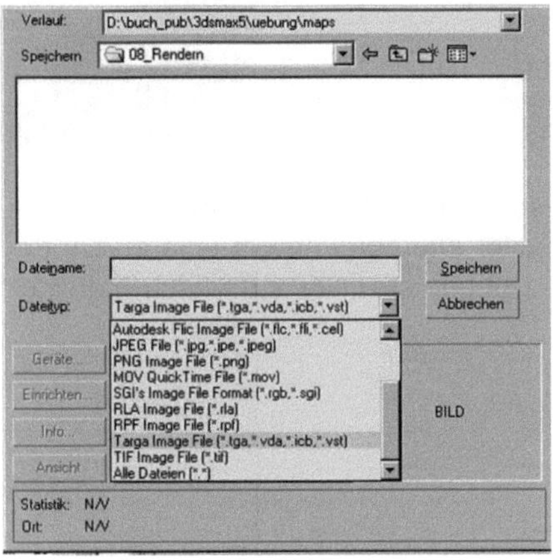


**Fig. 230.** Selection of the file format

If you want to generate, for instance, an image sequence from TGA files, you have to provide every single rendered image with a consecutive number. The pictures will then be labelled Picture0001.TGA, Picture0002.TGA, etc.

## Safe Frames

Safe frames will indicate the actual area to be rendered already in the viewport. Thus they provide an additional control for rendering. To activate this, click the right mouse button on one of the viewing window settings: *CONFIGURE • VIEWPORT CONFIGURATION • SAFE FRAMES.* Activate or deactivate via *CONFIGURE • SHOW SAFE FRAME.*

## Image Control with the RAM-Player

Once you have the scene more or less in the bag, with all settings done, and all that is left to do is to play around with various parameters, then this is the right moment to turn to the RAM-Player in 3ds max - an excellent tool for control. An example could look more or less like this:

You have finished the settings and rendered the first result. You could, of course, now save the file and open it with a picture viewing program like for instance ACDSee, or with a picture editing program like Photoshop, then render the next version, save it and compare both versions next to each other in the respective program. Or you could make use of the RAM-Player for direct and selective control on the spot.

**Channel A**

By rendering the picture without assigning a file output into the virtual frame puffer (Shift+Q), and then opening the RAM-Player (*RENDER• RAM-PLAYER*), the image can be loaded by selection in channel A via the option *OPEN LAST RENDERED IMAGE IN CHANNEL A*. The figure to the right shows the result in the RAM-Player. By now changing parameters and settings of, for instance, the light source, the changes can be tested directly.

**Fig. 231.** RAM-Player with the first image in channel A

**Channel B**

After the changes have been implemented, the new result can be loaded into the RAM-Player, which is still open. The new image is loaded into Channel B. In the figure to the right one can see the separator between the two results. Since the separator can be moved, the changes and their consequences can be controlled quickly and effectively. The only constraint is the fact that, for new comparisons, the pictures have to be rendered anew.

**Fig. 232.** RAM-Player with the both images in the channels A and B

## Increasing Efficiency

Apart from following textbook examples and having fun just playing around, there is usually an enormous pressure for time when producing stills or animations. Therefore it is important, especially for rendering, to use the available resources parsimoniously, and to accelerate the rendering procedure as far as possible.

A few remarks about a possible increase in efficiency at rendering:

**Model precision:** Be parsimonious with respect to too many details in your models. Include details only where they are really suitable. Always consider using a map for the representation. Especially beginners and CAD users have a tendency to model every detail. Every additional surface will cost rendering time.
**Resolution of Bitmaps used**: Normally bitmaps used for picture backgrounds or textures are taken from an original that was especially scanned for this purpose. Keep in mind not to use picture files with excessively high resolutions.
**Casting Shadows**: When installing the illumination, think about how many of your light sources will actually have to cast a shadow. Every shadow calculation will cost rendering time, no matter whether you are generating a Shadow Map during Scanline Rendering, or doing calculations for a Raytrace Shadow.
**Illumination Models**: Calculations for radiosity may often be advisable, but they are not always necessary. Always consider the possibility of getting by without this. If you want to use radiosity, keep in mind that the radiosity calculation for a scene with unmoving objects only has to be done once, whereas for a scene with moving objects, the entire light distribution has to be re-determined every time.

You may want to test the above criteria. For the same scene, take different amounts of polygons, to compare the effects of varying polygon counts. Use materials instead of models wherever possible. Try out various bitmap resolutions, comparing rendering times with a stopwatch. Play around with the shadow parameters of the light sources.

## Network Rendering

One fact certainly cannot be overlooked: rendering eats up time. The more complex the scenes, the longer the waiting times.

## General Conditions of Network Rendering

Time pressure, length of animation and available computing capacity, these are the general conditions which make a project feasible - or not feasible.

In an average example with a lot of geometry, some atmospheric effects and animated water, the general conditions could look more or less like this:

The average rendering time is approximately 3 minutes on a standard PC (AMD/INTEL 3200, 1,5 Gbyte RAM, WinXP). Supposing we want to generate an animation length of 1,5 minutes. The requirements are 25 frames per second (PAL, 720 x 576 pixels). This implies:

1,5 minutes x 60 seconds/minute x 25 frames/second x 3 minutes

and resulting in a rendering time of

6750 minutes or
112,5 hours or
4 days, 16 hours and 30 minutes.

It is difficult to image to source out the PC where you do your daily work for the duration of 4 days, 16 hours etc., only to get the calculation of one rendering job done.

Even if the calculation for the animation is restricted to the night (10 hours per night), it would take nearly two weeks to finish. A very long time.

Let us assume that there are 4 PCs in your office, which could all be involved in the calculations during a certain part of the daily working hours. This way the business could be done in a few days, and you would be able to administer your resources and especially your nerves, in a much better way.

## What does Network Rendering mean?

Network rendering is nothing other than having a certain number of PCs involved in the calculation of an animation. This explains why it makes sense to generate single frames: every network renderer receives a rendering order, and then calculates image after image. Each time an image is done, this is saved (e.g. as TGA), then there is a quick test, to see which picture is next, and on with the job.

In principle, one could think about outsourcing up the calculation for a video sequence like AVI or MOV, but this would not lead to any advantages worth mentioning. The big problem would always be the possibility of a crash. If the program stops the calculation for a movie sequence abruptly and unintentionally, the movie is usually useless, and you have to start over. In contrast, the calculation for single frames just carries on after the picture which was rendered last.

## Rendering Effects and Environment

A number of effects have already been introduced. Motion blur, depth of field, fog, atmospheric or lens effects - all of these can only be added when the calculation for the picture is done, after rendering. This way, a computer generated picture with clear structures and contours can be enriched with all kinds of mood effects.

**Fig. 233.** An Omni Light is applied for simulating the sun with the help of a Glow Effect

There are a number of effects for postproduction, which cannot always be clearly distinguished.

You have certainly come across some of these effects while working with picture editing programs. Effects can act in different ways on single objects (video post production), on the entire scene, or on light sources.

## Overview of Rendering and Environment Effects

**Table 9.** Rendering Effects

| Type of Effect | Description | Implications |
|---|---|---|
| Lens Effects | Lens f-stop effects which do not occur in nature, but can only be observed by looking through a camera. | Work only in connection with a light source. The light source is the starting point of the effect. |
| Blur, Soft Focus | Allows the later soft-focussing of the scene. | Optional on the entire picture, or on single components of the scene only. |
| Brightness/ Contrast | Allows the later setting of contrast and brightness of a picture. | Effects the entire rendered scene. |
| Depth of fields | Simulates the natural blurriness of components of the scene in the fore or background; originates in the camera lens. | Can be applied optionally on the basis of a camera and a target object, or on the entire scene. |
| Film Grain | Enables the simulation of film grain. | Affects the entire picture. |
| Motion Blur | Simulates the effect of a camera and its shutter speed. If movement occurs during the time the shutter is open, the image on film is blurred. | Affects the entire picture. |
| Fire Effect | Allows animated fire, smoke, and explosion effects. | For the application an atmospheric Gizmo is required to limit the extent of the effect. |
| Fog | Simulates fog and the atmospheric effect of color perspective (colors lose intensity with increasing distance). | Affects the entire scene. |
| Volume Fog | Allows fog effects of varying density within a defined enclosure. | Application of Volume Fog requires an atmospheric Gizmo. |
| Volume Light | Allows the generation of light effects based on light reflection with smoke or atmosphere. | Application requires a light source serving as starting object of the volume light. |

**Fig. 234.** By adding a Glow Effect to the light source, a pleasing result is achieved with little effort

## Layers for Post Production

It has become unthinkable to not work with layers for the postproduction of more complex projects. Working with layers makes life so much easier. Rendering results are given in layers or planes, which allow post processing in a composite or picture-editing program.

The simplest method for layer editing is the rendering of the respective layer as an independent image. The advantages of working with layers are, in short:

**Splitting of editing**: Large and complex scenes can be split up into separate components, and are rendered separately. This way, for example, complex moveable objects and the background are first rendered in separate image sequences, and later joined again in a composite. If the background needs to be changed, this can be done separately.
**Error checking**: When errors are detected in separately rendered scenes; it will in most cases be sufficient to re-generate the faulty component only.
**Speed**: A complex background (still) with cloud or similar effects can be rendered separately, and does not have to be rendered anew for every picture of the entire scene.
**Color matching**: This can be restricted to single layers, thus reducing costly keying.

### hint Alpha Channel

*One of the most important pre-conditions for the later composite of the visual material to be generated, are masked areas, e.g. which areas are to transparent, and which are not. The mask information can be saved in the Alpha Channel. Thus, an image format supporting Alpha Channels is a must. Potential formats are TIF, TGA, and PNG.*

In order to enable post-editing, the scene is to be rendered in layers. It has to be ensured that background changes can be done without too much effort. If necessary, effects like depth of field are to be added later. The color mood is to be adjustable. Furthermore, rendering time has to be shortened, because rendering the entire scene including all atmospheric effects will take too long. In the example the scene is split up into three layers, for demonstrating the basic procedure.

## Rendering in Layers

**Layer Background**

In the example the background consists of a color gradient. Supposing that the background will remain "still", e.g. there will neither be any changes in the cloud formations, nor in the glow effect, for this short sequence, then it is possible to generate a background picture remaining unchanged during eventual animations. By first deactivating all objects, and then rendering the image, the rendering result will show only the color gradient of the background. All objects have been deactivated. The glow effect remains active. TGA with 24 bit color depth is selected as output format.

**Fig. 235.** The scene with all components of the background

**Fig. 236.** The scene with volume fog, but without background, and without objects

**Layer Clouds**

The clouds in the scene were generated via volume fog, and are to be animated for a movie sequence.

Volumetric effects take a long time to calculate and later changes in the type of cloud movement may possibly be required. All objects in the scene, as well as the background, are deactivated before rendering. TGA with 24 bit color depth and additional Alpha Channel is selected as rendering format.

### Tip A practical Hint

*It makes life a lot easier if those objects of a scene, which are to be covered or blended in, are labelled with explicit selections, groups or the respective layers.*

### Layer Objects

All objects in the scene, the background with hills and mountain range, as well as the objects in the foreground, like boulders and grass, and the plain, remain unchanged. The background with glow effect, as well as the atmospheric effects of the volumetric clouds, is deactivated. TGA with 24 bit color depth and additional Alpha Channel is selected as rendering format.

Possible changes are limited to the background and clouds.

**Fig. 237.** The scene with all objects, but without background, and without atmosphere

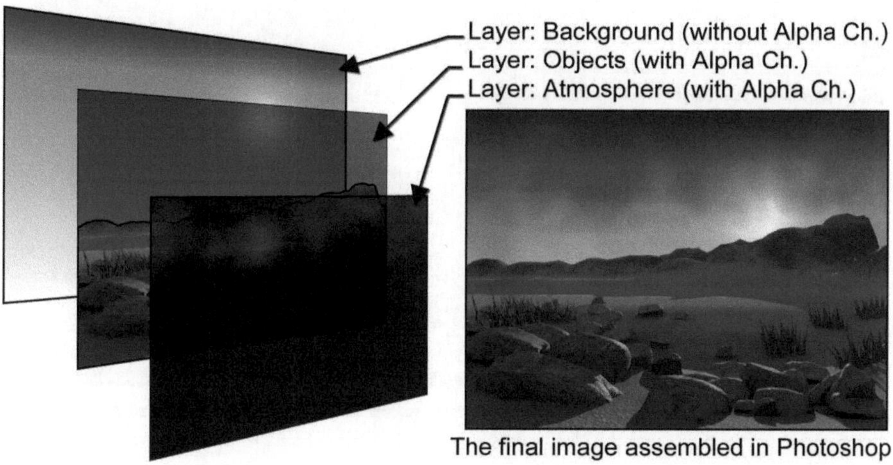


**Fig. 238.** The assembly of the finished renderings can now be done in any videopost or composite program

When planning to animate the atmosphere, the entire animation can only be done for the layer atmosphere (i.e. the volume clouds blended into the scene). It is important to add an Alpha Channel, which will contain the transparency information of the scene.

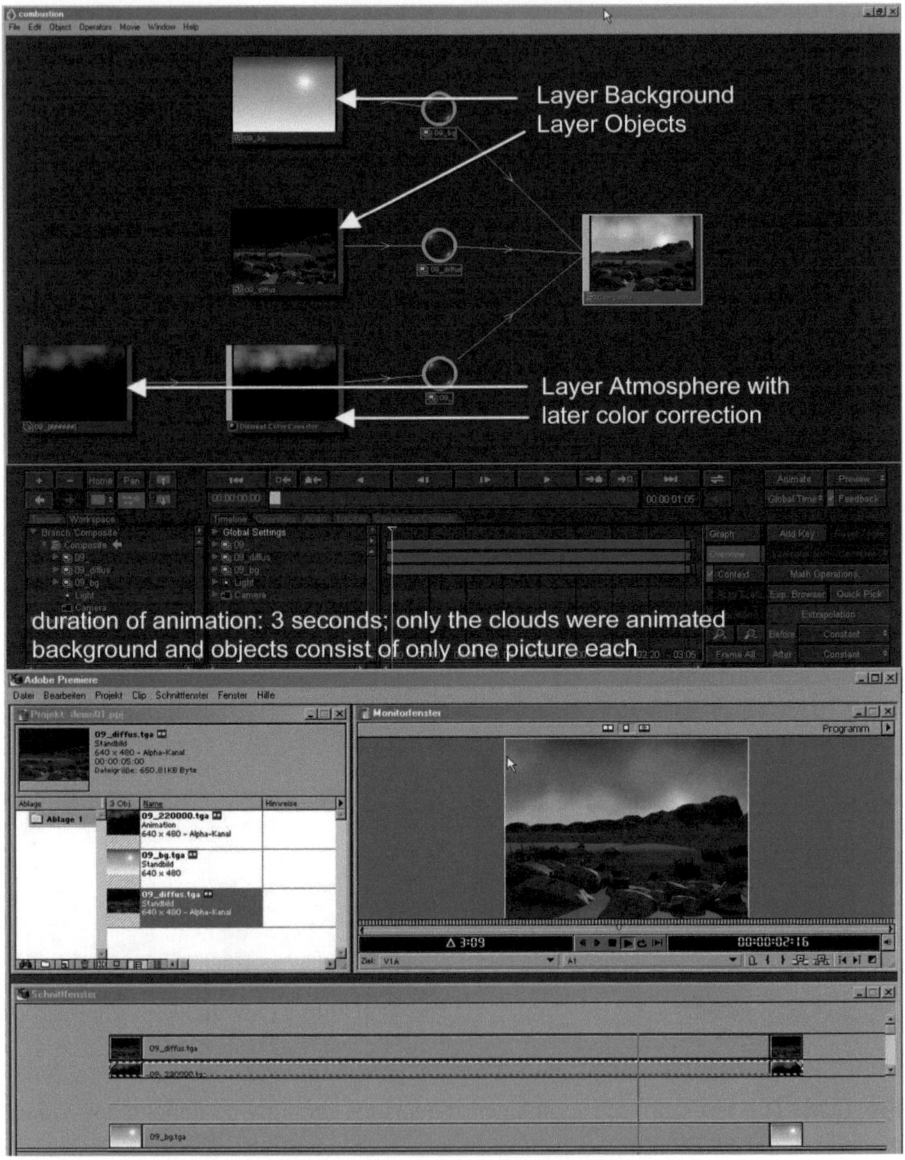


**Fig. 239.** The clouds, provided with an animation duration of 3 seconds, in the final Composite in Combustion (above), and Adobe Premiere (below)[5]

[5] Combustion is a program for video effects from Autodesk-Media, www.autodesk.com, Premiere the editing program from Adobe, www.adobe.com

## Using the Z-Buffer

You have generated an intricate scene which you want to edit with blur and, possibly, with color correction in Photoshop. The scene needs to be rendered in a very high resolution, and you know in advance that later changes to the settings for focal depth are inevitable.

The picture computation with the rendering effect focal depth, at such a high resolution, will not give a satisfactory result, and the computation with Multi-Pass, the depth of field of the camera, will require additional work and the cost of another 12 pictures (pre-settings), which have to be calculated as well. This will definitely take too long.

An alternative could be to do the post editing in Photoshop (or any other picture editing program, of course).

For generating focal depth, the information of the depth of a scene is needed, i.e. the information about the distance of an object from the camera. This information is included in the so-called Z Depth. Similar to the Alpha Channel for masking objects, Z Depth provides a gray scale picture. A specific distance to the camera is assigned to each gray scale value. White in this case means: in direct proximity to the camera, while black pixels are an indication for maximum distance from the camera.

With the help of Z Depth it is possible to assign, and just as important, to change, focal depth quickly and easily in Photoshop.

### Z Depth

The file output is to be generated as a TGA format. In the rendering dialog (F10) under *RENDER ELEMENTS* the option *Z Depth* is added. Under PARAMETER SELECTED OBJECTS FILES the name for the file that is to be generated is filled in automatically: Name_of_File_Z Depth.TGA.

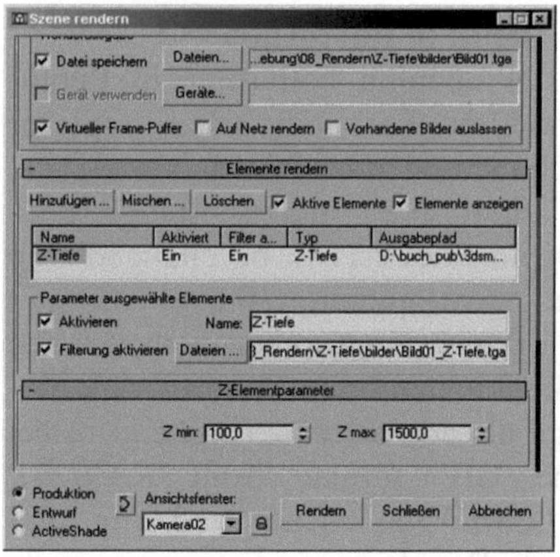


**Fig. 240.** Render Elements and Z Depth

**Fig. 241.** The finished image (left) and the Z Depth information (right)

### hint Z-Element Parameter – keeping the distance

*When rendering in a Z Depth file you have to watch out for the dimensions within the scene. Before rendering it is recommended to measure (CREATE PANEL HELPERS TAPE) the distance of the camera to the (for the edition relevant) most distant point from the camera, which is still to be part of the Z Depth. You can subsequently enter these values under Z-ELEMENT • PARAMETERS as Z MIN and Z MAX.*

**Z Depth in Photoshop**

After opening both files in Photoshop, there are only a few steps for achieving depth of field.

First the image Image01_Z-Depth.TGA has to be changed to a gray scale picture: *IMAGE • MODE • GRAY SCALE*. (Another method is the direct generation of a gray scale image already at rendering, e.g. select the GIF format from *RENDER ELEMENTS* in the rendering dialog). By changing to a gray scale mode, a new channel is generated in Photoshop. Now duplicate the background layer of the entire image, labeling the newly generated layer "Blur".

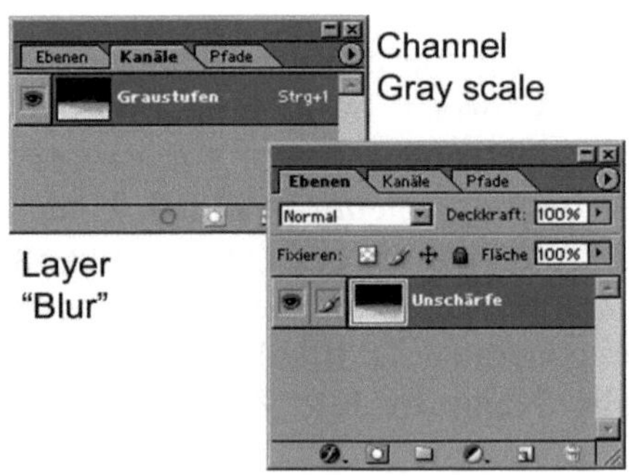


**Fig. 242.** Channel "Gray scale" and Layer „Blur“

When selecting the option *SELECT • LOAD SELECTION....* from the menu, the channel “Gray scale” from the gray scale file can be loaded as selection.

In the menu *SELECT* choose *INVERSE*, an then in the menu *FILTER BLUR* select the filter *GAUSSIAN BLUR.....*

**Fig. 243.** Channel and Layer in Photoshop after changing the image to a gray scale image

**Fig. 244.** The finished scene in Photoshop with exaggerated Gaussian Blur

## Rendered Images and Office Products

So far, we have been using terms like video post editing composite, image formats and compression procedures, and hopefully some of the parameters for rendering data from a 3D environment may have become a little clearer, but there still remains the question of which format to use for which purpose.

Therefore, briefly, a few suggestions and recommendations for possible areas of application and special requirements, or, in other words, Where does what happen with which programs and images?

## Integrating Images in Office Documents

Expert's reports and statements represent a wide range of application for visual data from various fields. There is a different approach, depending on whether the visual data will only be used "internally" (in-house with the printing being done on the laser printer), or whether they are destined for the print media (publications, advertisements).

When, for instance, images are embedded in WinWord, in other words, they are imported, and then these are adapted, during import, automatically into Microsoft's BMP format. The use of memory-saving JPEG-files does not make sense in this case.

The resolution for images for internal use is normally quite adequate at 150 dpi. For graphic material and pictures requiring a high precision with regard to details, and which are destined for use in the print media, the resolution can be increased to 300 dpi for color pictures and to 600 dpi for gray scale graphics.

The general recommendation: do not import visual data, but instead preferably link it. The problem here is that the external images (preferably saved in a separate file labeled pictures) have to be taken into account when copying. However, the advantage is obvious, for the images remain available for other uses.

Recommended picture format: TIFF or EPS.

**Table 10.** Images in documents for internal use or for printing

| Image data | Internal use | Print |
|---|---|---|
| Image Type | (BMP,) TIFF (RGB) | TIFF, EPS (CMYK) |
| Integrated into document | No | No |
| As separate (linked) file | Yes | Yes |
| Resolution | 150 dpi | 300 dpi |

## PowerPoint Presentations

A very popular and much used tool for the presentation of data and contents is Microsoft's PowerPoint. As a rule, the visual material is projected onto a projection surface via a beamer, in a normal resolution of 1024 x 768 (1280x1024) pixels, and in varying light intensities. Therefore picture data can safely be limited to 96 dpi.

Movie sequences, too, can be integrated into PowerPoint. In this case, the classic maximum resolution of 4:3/ 640 x 480 pixels is quite sufficient.

It is important to ensure that the Codec used for the movie has been installed on the computer used for the presentation, or, in the case of a MOV, that a Quicktime-Player is available. In any case, preference should be given to AVI movies, because these cause the least problems when working with PowerPoint.

**Table 11.** Images in documents - PowerPoint

| Image Data | |
|---|---|
| Image Type | JPG, BMP, TIFF (RGB) |
| Integrated into document | Yes |
| As separate (linked) file | No |
| Resolution | 96 dpi |
| **Video Data** | |
| Image Type | AVI, MOV |
| Integrated in document | No |
| As separate (linked) file | Yes |
| Resolution | Max. 640 x 480 |

## Web Publishing and Digital Documentation

Publications and representations in an Intranet and the Internet require considerably more attention with respect to file size. Here, the picture formats JPEG, PNG and GIF should be used. The respective programs for meeting these requirements are, for instance, Dreamweaver, Frontpage, or GoLive.

**Table 12.** Image

| Image Data | |
|---|---|
| Image Type | JPG, PNG, GIF (RGB) |
| Integrated into document | No |
| As separate (linked) file | Yes |
| Resolution | 72/96 dpi |

## Summary

Rendering, image formats, and post-production - these were the subjects of this chapter. The current image formats suitable for 3D presentations were introduced, and questions were asked about which image format is best for which purpose.

Compression methods for still frames and their respective advantages and disadvantages were dealt with. It should be known, by now that images are available in a compressed or an uncompressed form.

Compressing image data has the advantage of saving memory. But these savings are always at the cost of image quality.

When rendering 3D animations, it makes sense to first render a high quality, lossless picture format. These sequences of single pictures can later be converted according to specific demands.

You are now familiar with the two preferred video formats AVI and MOV, and you know that these co-called container formats can include different formats. These Codecs are significantly responsible for the file size and quality of the film sequence.

Taking 3ds max as an example, some significant aspects for dealing with the rendering procedure were highlighted.

The RAM-Player in 3ds max, for example, is a very useful tool for doing a quick check of variable parameters of single frames and movies.

There are, particularly in rendering, methods to optimize and to speed up the rendering process. You have come to realize that model precision, resolution of bitmaps used, shadow characteristics, and the selection of a suitable model for illumination are all important factors for the speed of image computation.

Furthermore, some rendering effects were discussed. Rendering effects labeled effects, such as Lens Effects Glow, are separate domains of the rendering process.

Fog, volume fog, and fire effects are all atmospheric effects that can be found in the rendering environment.

You also know that the option Render Elements gives you the possibility to save specific components of an image, like for instance Z Depth, in a separate file, and to use this, for example, in a picture editing software like Photoshop, for compositing purposes.

# Interaction with 3D Data

*A short overview of the options for interaction with digital terrain models and landscapes.*

In the age of Half Life 2 and other games with an enormous modeling effort, the age of Google Earth and TerrainView-Globe, fast computers, cheap beamers, and fast Internet connections, the subject of interaction with 3D data is booming.

Data interfaces between various program software packages function (more or less) well, and the times for expensive real time 3D graphics of various supercomputer servers are (nearly) over.

In short: the interactive viewing of 3-dimensional data is up-to-date. It makes sense, it facilitates comprehension, it opens up new ways of understanding, it is feasible, and it is a lot more fun than viewing movies or stills in a given (linear) form.

## Interaction?

The interaction with data in real time is nothing new, really. Already more than a decade ago, it was possible to generate Real Time visualizations by various methods; one only has to think of VRML (Virtual Reality Modeling Language), and its much looked after and highly praised successor X3D. In the meantime, quite a few providers are competing in the field of interactive Real Time representation of 3D data, often with very good commercial solutions for the demands of "accessible worlds".

However, most of these software packages still have one thing in common – the problem of having been able to handle large "sets of data". Some of these tools may be excellent for product visualization, but their resources are often insufficient for visualizing a terrain, or for representing large, complex scenes.

While it is relatively easy to integrate scenes made up of simple geometries into a Real Time environment, it is a different story altogether when you are faced with a complex large mesh. When this has to be fitted additionally with collision control, the fun soon ends and the performance will break down.

Another problem is the exchange of 3D data. Limitations in pure geometrical exchange formats will often result in the chosen format not being able to handle the required amount of data.

One example for this is the format 3DS, which has a limitation regarding the number of polygons in the file.

The tools specialized for the demands of terrain modeling are a different matter altogether.

## General Demands on Real Time Representations

The performance of graphics and games of the latest generation have opened up many new possibilities which have eagerly been embraced by all of us. New standards were set, but the question remains, which criterion is actually at the top of the list when regarding the visualization of terrains.

Below is a list of the most important aspects for a representation of digital terrain data:

Representation of unchanged geometry
LOD – Level of Detail
Integration of "large" textures
Speed
Action(s) / Behavior
Handling / Navigation
Platform / Presentation
Data transfer

These criteria refer to the viewing of, respectively, the navigation in digital terrain models (DTM) and the interaction with these data in a VR environment. The additional use of provision for author input, like for instance in a game developing environment, is not relevant here.

### Representation of Unchanged Geometry

Independent of which specific terrain model is used, one of the most important requirements is the option of being able to view geometrical data in the original state. This is important particularly for Triangulated Irregular Networks (TIN), but also for all other Grid DTMs.

When exporting from different software packages, many data exporters for interactive viewers or VR environments will automatically reduce the geometry topography according to specific criteria, by means of built-in optimization algorithms. Although this function will effectively reduce the file size, it may lead to undesired results and to loss of information.

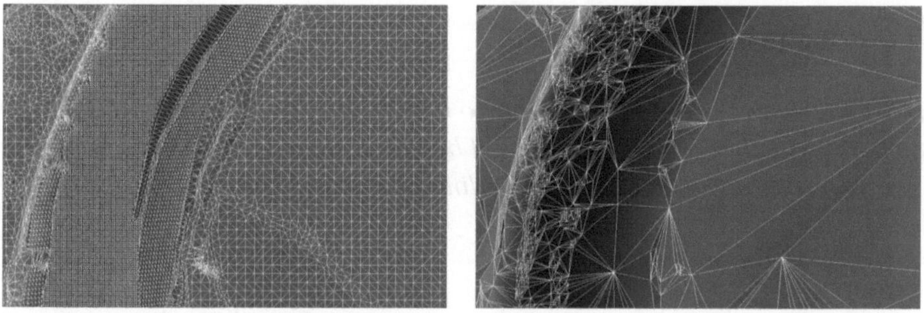


**Fig. 245.** The picture on the left shows the original set of data, in the picture on the right these were reduced in favor of the file size – by removing important information

## Level of Detail (LOD)

Level of Detail allows the dynamic representation of data taking the part of the scene you are looking at, and the distance. Here, the representation of objects and textures becomes increasingly simplified, with increasing distance from the camera -"simplified" meaning the automatic reduction of geometry and texture, while ensuring that the objects can still be identified at any time. The closer the object is positioned to the camera, the more detailed it is represented, until, at a specific distance to the camera, it loses all of its geometry and texture information.

## Integrating "Large" Textures

Textures ensure a suitably realistic and/or informative generation of 3D-data. Two types of textures for interactive use are of special importance:

Textures generated in any 3D program (e.g. the import of light/illumination information, as generated, for instance, by Texture-Baking),

Geo-referenced image data, e.g. aerial photographs.

The textures integrated into the VR environment should, independently of their size, be represented quickly and as losslessly as possible.

**Tip Two to the power of ...**

*Most interactive programs are able to handle picture sizes very well that contain a number of pixels expressed by two to the power of n ($2^2$, $2^4$, $2^{16}$). This fact also results in a corresponding increase in speed.*

## Speed

In any case, the software should be able to allow, without extra effort, a flowing movement (Flythrough, Walkthrough) through the topography. In other words, the rendition of the image should be done without shaking, vibrations, or loss of frames.

To achieve the impression of "real" navigation within a 3D environment, a repetition rate of about 10-15 frames per second is recommended.

## Behavior / Actions

When navigating through a 3D environment, it makes sense to be able to trigger specific actions, or to be able to react to such actions with a specific behavior.

The integration of a collision control should always be included, as nothing will be more annoying than the uncontrolled penetration of polygons, and the sudden view of the topography "from behind or from below".

Furthermore, it will give you peace of mind, particularly for navigational purposes, to integrate so-called Hotspots or POI (Points of Interest) into a scene or an environment, and to provide these with additional information or links. Additional information can be, for instance, 3-dimensional texts, objects, or additionally integrated image data. Links can be targets inside (camera view to camera view), or outside the environment (e.g. HTML-link).

## Handling/Navigation

First of all, it should be easy to find your way in familiar surroundings, i.e. in your operating system. The easier the navigation for the user, the easier the handling is, and the sooner the environment is accepted.

It is important to be able to do the navigation with standard methods of input, like a keyboard and mouse.

Apart from this, it is of great importance to be able to transfer the fourth dimensional component (time) with sufficient ease, as normally the monitor acts as a window to the VR environment, when navigating through the 3-dimensional space. Some relevant features for an easy navigation are:

Movement along all three axes of space
Viewing of object (Exploration)
Pedestrian (Walkthrough) or vehicle (Drive) mode allowing the taking into account of gravitation, and
Flight mode
Changing speed

Furthermore, additional elements for a straightforward handling of specific events are desirable. The navigation should be feasible with the current input options of a standard PC. These are the keyboard and the mouse. An optimal integration of other input tools, like for instance a joystick, is "nice to have".

Less is more! It is better to have a fast, functional, and easy to handle surface, than too much information and complicated handling.

## Platform and Price Politics

If the VR environment is linked to a specific product, one should find out beforehand whether, for the visualization of the data, this product needs to be installed on a presentation computer, or whether it is possible to use a Plug-in or an independent viewer. Here are some points supporting the interactive use of the platform:

Separate viewer, working independently of the original product - best without costly installations
Platform comprehensive (Windows, Linux, Apple)
Browser integration via Plug-in (Plug-In possibly for Windows, Linux, Apple)

Imagine you had to pay a fee to the software producer for every CAD file generated to aid the exchange with project partners – it's unthinkable, isn't it?

But unfortunately this happens very often. Web-based solutions with a complicated and costly licensing structure are often difficult to handle, and expensive too.

In the best case, a once-only product price is paid, which includes the publication or the use of the interactive scenes generated. All viewers or plug-ins should be available free of charge.

### Data Transfer

Data transfer and an uncomplicated rendition of existing 3D information are decisive factors for either VR or interaction. The simplicity of the interface to existing 3D tools, be it 3ds max, Cinema, Maya, or one of the numerous other products, will decide about acceptance and use of interactive environments.

Some points for data transfer:

Integration into existing 3D tools (Plug-In) or

Data transport which takes over the existing information in a "clean" fashion. This means: no postproduction or editing will be necessary!

All the necessary options for viewing a 3D scenario should already be included and available in the 3D tool. The need for a costly editing later on will act as a deterrent.

## Procedures and Methods

The interesting thing about VR, interaction, and Real Time is the fact that, for over a decade, nothing has really changed in the fundamental methods of data representation.

Programs have become more sophisticated, hardware speed has increased, and mice and joysticks have become more sensitive and reacting a lot faster.

But still the main part of visualization of Real Time applications is done on the monitor. Large projection surfaces still have not become a standard. The input tools most widely used are still the mouse, the keyboard, and maybe a joystick or a space mouse. Stereoscopy has still not gained much ground, even though the production cost for shutter-glasses or glasses with polarization filters is now only a small percentage of what it was about 10 years ago. Neither the data glove nor force feedback of input tools have been able to conquer the market. The head-mounted Display has disappeared altogether.

What remains is the usual environment. This is composed of a monitor and (maybe) a beamer, a mouse and a keyboard.

The common methods are:
Interaction with image data (Panorama) and
Interaction with geometrical data.

Examples for these are QuickTime VR, data viewing with VRML, and a range of products from various commercial suppliers, suitable for use in different fields.

## Interaction with Image Data

"Interaction with image data" refers to the finished, rendered picture file which allows an all-round view of a 3D scene, depending on the method of rendering. In this special case the final result is presented in high quality. Neither is it possible to move through, nor can a "genuine" interaction with the objects of the scene be achieved. But for many presentations this will be quite sufficient. The one tool, which has been able to conquer a niche and sustained its position is the QuickTime Player with its possibility of representing QuickTime VR data or in other words, panorama pictures.

### Quicktime VR?

Quicktime VR is a technology developed by Apple, which allows limited interaction with panorama views or 3D objects. The basis for 3D visualization with QuickTime VR is a panorama picture in the form of the projection of an object onto a sphere or a square, as a pixel file. This projection of a 3D environment is first generated, and then edited with the suitable software; it manages to convey the impression of 3 dimensionality.

The unbeatable advantage of methods like panorama pictures or other procedures as offered by QuickTime VR is undoubtedly the possibility of generating photo-realistic representations.

Where the rendition of a "genuine" 3D model is needed, like in a game environment, the limitations are soon found, due to the large amounts of polygons.

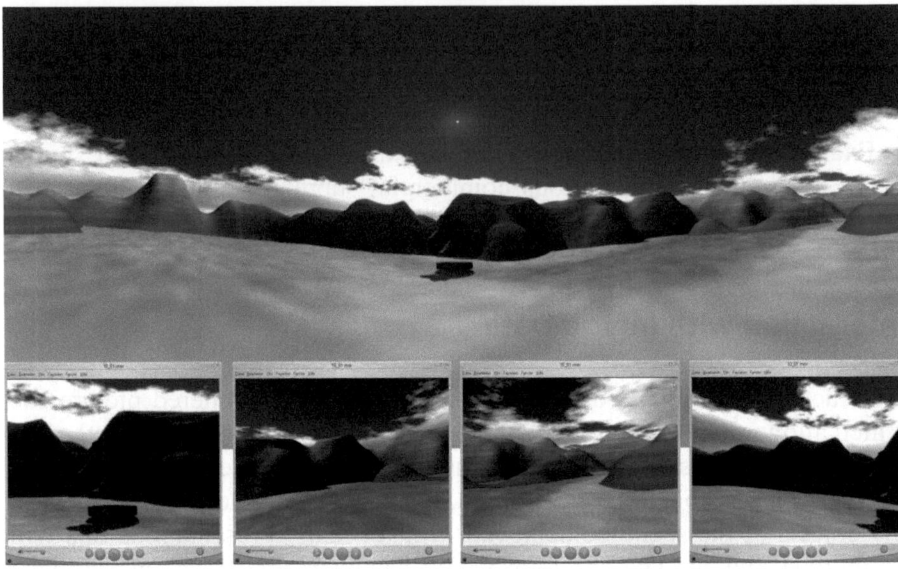

**Fig. 246.** Export of a scene as a panorama picture – the figure shows the projected result as a single picture (above), and various settings in the QuickTime Player.

Although for viewing QuickTime VR data, the QuickTime Player is needed, it is so widely used by now that it can be considered a "semi-standard".

The interesting thing when using QuickTime VR is the fact that the result can be rendered directly from the application. A special rendering of the scene, as is the case in 3D Real Time software, is not necessary.

For nearly all 3D applications, there is now a suitable export option available for generating such panorama pictures. In 3ds max, for instance, this can be found under *RENDERING* • *PANORAMA EXPORTER*.

## Interaction with Geometrical Data

This means the representation of a 3D scene in Real Time. Here, the problem lies in representing as many geometry and texture information details as resource savingly as possible.

## Preparation

Before tackling the data export of models from 3D programs into any 3D author environment, it is usually necessary to do some preparation concerning the geometrical information and the textures.

It is recommended to remove complex modifiers used for the geometry, i.e. to first transform the 3D objects into a workable mesh or polygon. It also pays to take a quick look at the pivot point as well as at the fitting of the direction of the local coordinates of the object to the global coordinates.

The rule for textures is: procedurally generated textures cannot (normally) be transposed for an interactive representation. Therefore it is necessary to first generate a material including a diffuse channel equipped with a bitmap. Aerial pictures are generally available as pixel graphics, and thus do not present any difficulties.

An existing procedural map could first be rendered into a file via Texture Baking, then assigned as a bitmap to the diffuse channel of a new material, before tackling the data export. In case “holes” were omitted by reverse normal alignment using a double-faced material in the 3D program, these faulty areas should be corrected before export. Such holes are likely to cause problems during texture baking.

**Fig. 247.** A random topology with different textures – the material editor to the left with the original surface with Gradient Ramp, and to the right with the bitmap generated in the diffuse color channel via “Render to Texture”.

The following data formats are suitable as texture data formats for most 3D Real Time applications:

Joint Photographic Experts Group (*.jpeg;*.jpg)
CompuServe Graphics Interchange (*.gif)
Data Dictionary System (*.dds)
Portable Network Graphics (*.png)
SGI Image Format (*.rgb)
Tagged Image File Format (*.tif;*.tiff)
Truevision Targa (*.tga)
Windows Bitmap (*.bmp).

Suitable data exchange formats for geometries are:

OpenFlight Files (*.flt)
Autodesk 3D Studio Files (*.3ds)
Virtual Reality Modeling Language VRML 1.0/VRML97 Files (*.wrl)
Wavefront Files (*.obj)
LightWave Object Files (*.lwo)
Open Scene Graph Files (*.osg).

If you happen to own Polytrans from Okino, or Deep Exploration from Right Hemisphere, you will be able to convert various 3D formats from any 3D application.

### *Reduction of Geometry*

In the course of normal work, you very often come up with complex constructions of modifiers and distortions, until you finally get to the desired result. First, for instance, a noise is assigned to a specific terrain, which is subsequently provided with softer contours by a smooth modifier, and afterwards this is followed by an adjustment of the UVW coordinates of the object, etc. Normally, these distortions and modifications remain intact in the original scene; this allows quick adjustments or changes later on. However, when used in an interactive environment in a different program, this does not make sense and, at the latest during export of the geometry, all remaining scene modifiers will be reduced to a workable mesh or polygon. This automatism during export makes a lot of sense, but unfortunately sometimes strange effects are caused because of this. In order to avoid these potential uncertainties, geometries should be manually simplified before export.

For this, it is best to make a copy of the scene and save it under an explicit name (*NAME OF SCENE_INTERACTIVE.XYZ* ...).

Although one of the premises for the interactive presentation of terrain data is as precise a rendition of the model as possible, on occasion the generation of the DTM has a more qualitative character. In such a case the correct rendition of all DTM elements is not absolutely compelling, and there is the possibility of simplifying the topology of the model. This simplification implies the reduction of polygons and can be achieved by various methods and tools. In 3ds max, for instance, there is a choice between the modifiers *OPTIMIZE* or *MULTIRES*. Both modifiers remove polygons, but according to different criteria. Often a surface displaying the same height throughout, originating, for instance, from a set of raster data, will be described by a few points only. Points of identical height, located within this surface, are simply removed.

The procedure for reducing constructive and optimizing modifiers is described below.

**Modifiers**

A random box is converted into an editable polygon. Then the terrain is distorted by a *DISPLACEMENT-MODIFIER*.

A gray scale file, generated in Photoshop, is used for height information. Following this, a *UVW-MAPPING-MODIFIER* is applied, to adjust the alignment of the texture (a procedural color gradient). This construction has the advantage that the *DISPLACEMENT-MODIFIER* as well as the UVW coordinates can be easily changed, at any time.

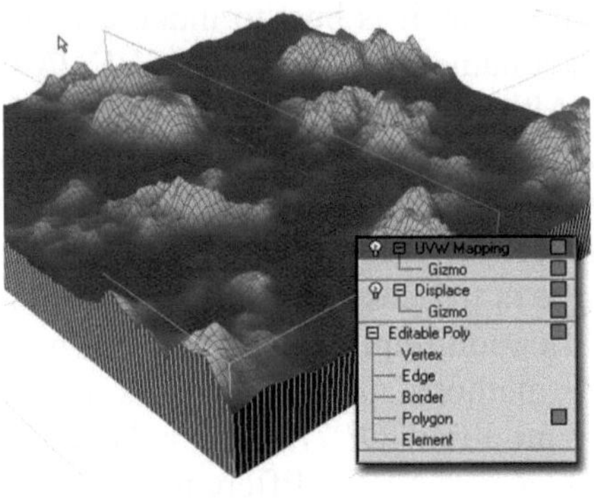


**Fig. 248.** „Original Terrain“ with all available modifiers

**Collapsing**

Now all modifiers are assigned to the object and fixed to the geometry. A later manipulation of separate modifiers will not be possible any more. One click of the right hand mouse button on the open modifiers opens a window in 3ds max which allows to “collapse all” contents. The result is optionally an editable mesh or polygon.

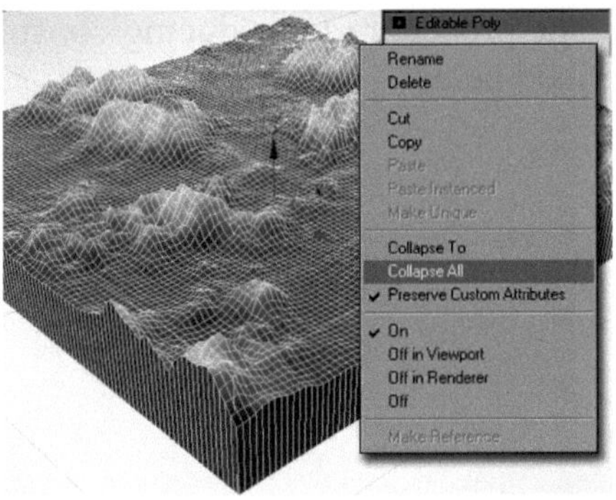


**Fig. 249.** Adjusted terrain – all existing modifiers have been collapsed. What is left is an editable polygon.

**Reducing**

The possibility of simplifying a mesh, of optimizing it, will lead to a further, immense reduction of the data volume. However, extreme caution is advised here. Depending on the purpose of the model, it may not be wise to optimize the mesh at this late stage - especially when the visualization is to be used for discussion, and is to serve as a decision aid.

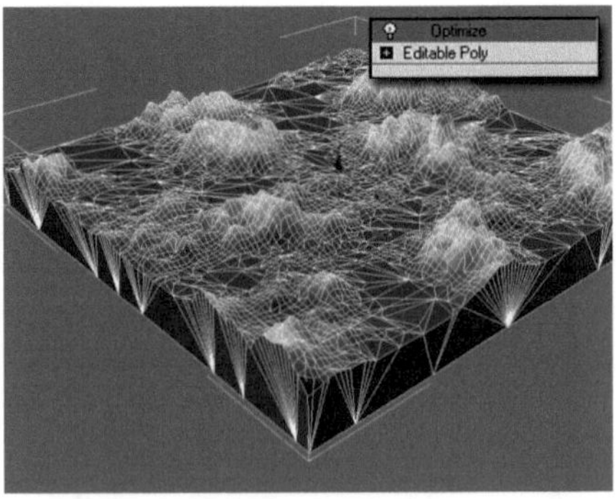


**Fig. 250.** Optimized terrain

## *Texture Baking*

After the geometry has been finalized and, if necessary, additionally optimized, the object material has to be "baked" into a texture. Texture Baking is nothing other than assigning light/illumination and shadow information to the material of the object, and to export this, together with all other characteristics, into a map.

The reason for this is that, although many Real Time applications are able to represent an object with its texture (provided this is a pixel graphic) as well as illuminate it, they cannot calculate the shadows.

Consider the visualization including a radiosity calculation, which you want to represent in Real Time. Here the illumination information can be "welded" into the texture, which allows, even in Real Time, a simulation with radiosity calculation baked into the texture.

### Materials

A multi material has been assigned to the DTM. A blue surface filling provides the color for the frame, and a color gradient is used for the height-coded representation.

In order to demonstrate the effect of the illumination combined with the corresponding shadows, three cubes with a checker material were added.

The sub-materials are made up of procedural maps, which normally cannot be integrated into a Real Time environment.

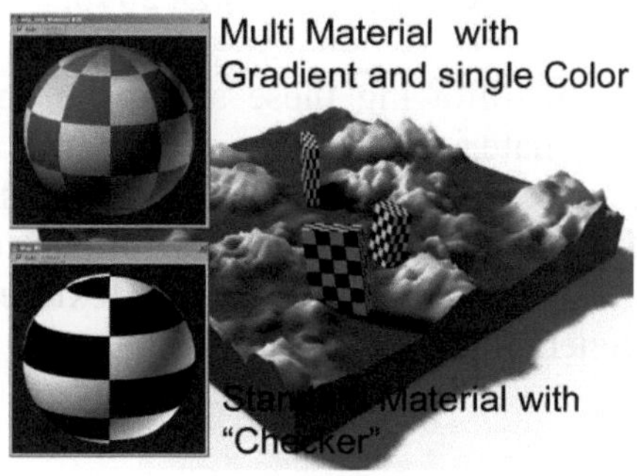


**Fig. 251.** Multi Material

**Render To Texture**

By selecting *RENDERING • RENDER TO TEXTURE,* a dialog for specifying the required settings is opened. By taking care of the following settings:

1. Define *OUTPUT-PATH.* This is where the results will be rendered.
2. *MAPPING CO-ORDINATES-OBJECT: USE AUTOMATIC UNWRAP, CHANNEL* 3
3. Activate *BAKED MATERIAL • SAVE SOURCE • CREATE NEW BAKED.*

Choose *OUTPUT - ALL SELECTED,* click *ADD* and select *COMPLETE MAP* and activate *TARGET MAP SLOT: DIFFUSE COLOR, MAP SIZE 2048.*

By following these steps, the entire material can be rendered into a texture. It is recommended to choose JPG or PNG as the output format, to ensure the versatile use of textures generated in this way.

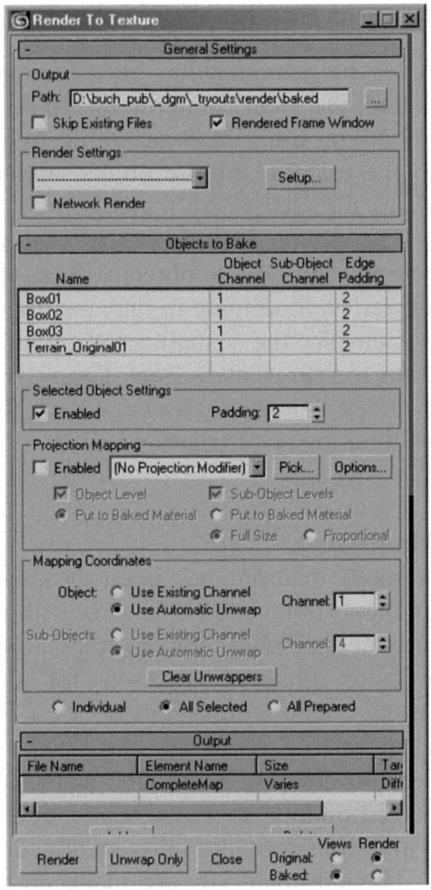


**Fig. 252.** Render To Texture Screenshot

**Assigning Textures**

Next, a new standard material has to be generated for each object. The rendered texture result is assigned to this standard material, in the diffuse color channel, and the map channel is set on 3. The newly generated material is then mapped onto the object.

In 3ds max, the modifier *AUTOMATIC FLATTEN UVs* was additionally assigned to the various objects by the procedure *RENDER TO TEXTURE.* Selecting the object and clicking on the right hand mouse button the Quad open menu. Select "*CONVERT TO EDITABLE POLY OR MESH*".

After converting, the UVW information is directly assigned to the object. In the field of interaction there are many programs, which have difficulties dealing with different channels. TerrainView, for instance, only supports map channel 1. Therefore "baked" information from map channel 3 has to be copied into map channel 1. This is done by *TOOLS • CHANNEL INFO*. The map channels 2 and 3 can be deleted after copying. In the material editor the map channel has to be reset to 1, and then nothing will keep you from exporting.
As a quick check to find out whether the textures have been assigned without any faults, an export as VRML file and testing in the browser with the VRML Viewer can be very helpful.

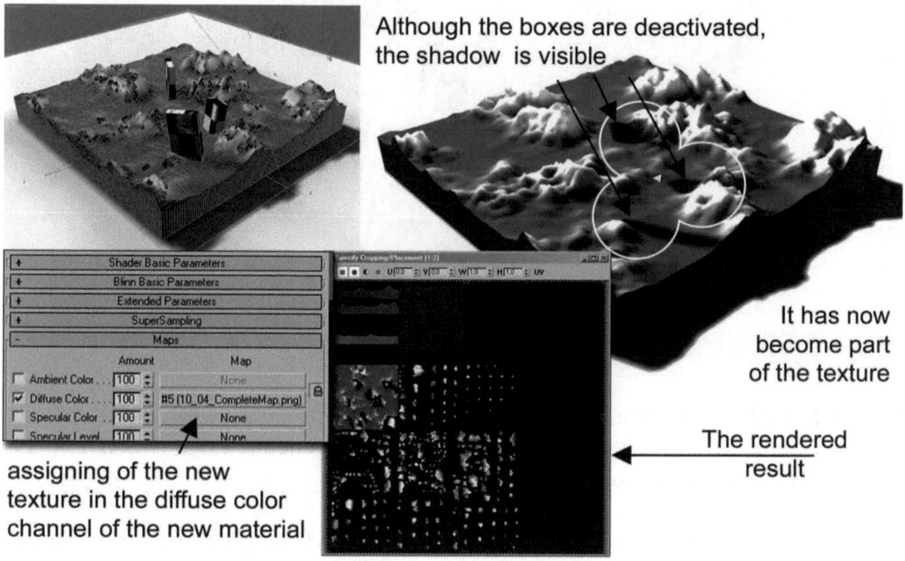


**Fig. 253.** The material information now "baked" into a new bitmap not only contains all former information of the multi map but also the illumination information including shadow data.

## VRML

The use of VRML is a very reliable as well as a very sensible way for achieving 3D interaction with 3D data.

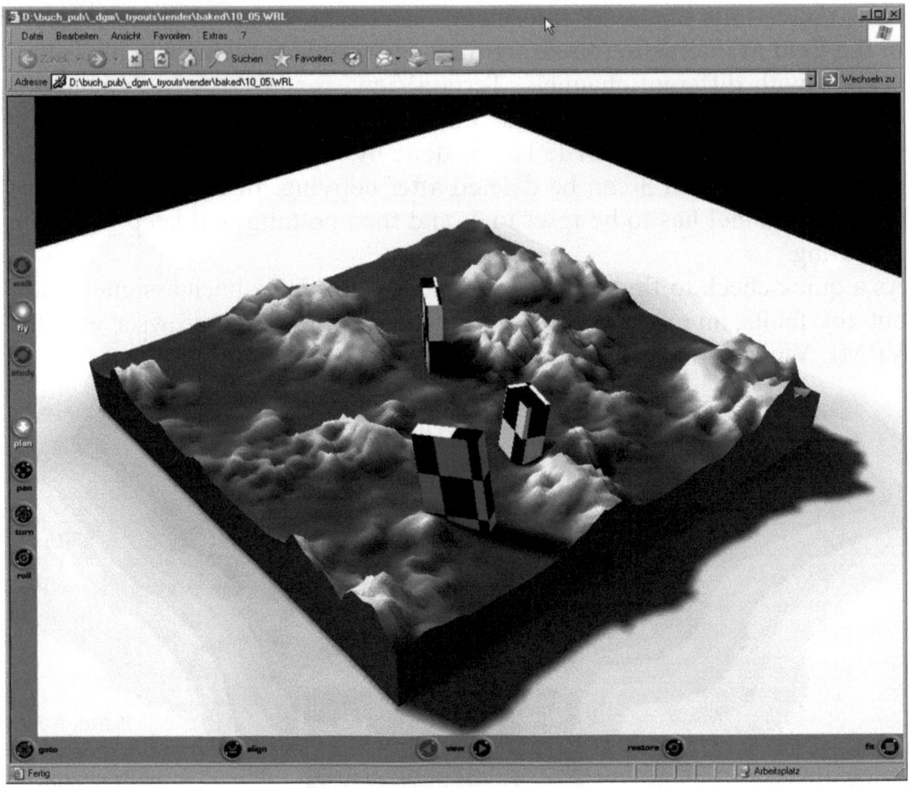


**Fig. 254.** VRML for checking: the terrain in the Cortona Viewer[1] in the Internet Explorer, after it has been assigned the textures generated via RENDER TO TEXTURE, and after export as a WRL file.

VRML[2], **V**irtual **R**eality **M**odeling (also called **M**arkup) **L**anguage is a descriptive language for the representation of 3D objects and scenes. VRML is a language in ASCII format, which can either be integrated into standard web browsers via suitable Plug-Ins, or viewed with the help of Standalone products.

The advantage of such an interpretative language is its independence of specific platforms, and its availability in the World Wide Web. So-called Hyperlinks can be integrated into VRML as well as into HTML data.

---

[1] http://www.parallelgraphics.com/

[2] You will find the complete VRML97 specification under http://www.vrml.org/Specifications/VRML97. Furthermore, the document contains technical data with respect to the behavior of exported VRML97 worlds.

Selecting activates the hyperlinks i.e. by clicking on a highlighted object, thus allowing the integration of information within the environment of the Intranet or the Internet.

A decisive disadvantage, however, is the fact that the description of more complex models may contain a large amount of data. When using the Internet, this implies long transmission times. In this case it is essential to optimize the data in such a way that a sufficient data rate will be sustained.

When the use of the VRML format is NOT intended for online use, then VRML is still an excellent alternative for the interactive checking of 3D data in Real Time. However, this interaction should be restricted to the viewing of topologies and simple primitives. Attempting to transpose polygonal plants in a VRML scene will soon lead to a dead stop.

When a rough representation is needed for online use, it makes sense to stick to the following:

Use GIF, JPG, or PNG as the file format. There is an automatic PNG support integrated in most VRML viewers, which makes this format preferable.

Keep an eye on the number of polygons. The smaller the number of polygons to be dealt with, the higher the performance will be. 100.000 to 300.000 polygons are a quantity that can be administered and represented without problems when using modern graphic cards (with a minimum of 128 MBytes).

LOD: Level of Detail is an option which can be defined in your VRML file, and by which you can control the detail precision of your objects in relation to their distance to the camera.

Use instances instead of copying objects. This will also reduce the file size tremendously.

Reduce the representation in your Web browser, so that it does not take up the entire monitor space. Define a window for the integration of your VRML file. You will achieve this via the integration order:

```
<embed SRC=test.wrl WIDTH=320 HEIGHT=240>
```

this will open a window measuring 320x240 pixels for showing the VRML file on the HTML page.

Always define at least one camera and one light source.

## 3D Authoring Applications

Authoring applications such as Anark Studio[3], Quest3D[4] or Virtools[5], to name only a few, are, actually, the "crème de la crème". These tools originating from the field of product visualization or game development will allow you to do more or less everything that makes the 3D interactive heart beat faster. You can generate entire game levels as well as visualize simple product representations of excellent quality.

High quality has its price, however. Drag and Drop won't do the job, scriptwriting and programming know-how are essential here. The simple use of a defined surface is not enough, unfortunately.

Data interfaces are included in most current 3D programs, and data exchange is seldom a problem.

The preparation for the scenes as described under the heading "Reduction of Geometries" is a necessity, just like the in-depth familiarization with the respective user interface. It takes a long time to get into the program and you should only consider trying to become more familiar with such a tool when you expect to use it frequently (or if it gives you such a lot of pleasure that you see it as a hobby).

Once the rough geometry has been imported, there are no limits to the imagination. Everything is possible, from the collision control to events right up to atmospheric effects in Real Time.

Another advantage of authoring applications lies in the fact that current graphic cards are used to the limit. Independent physics engines are in general use, and very realistic shadows are supplementing entire representations of natural landscapes and plants.

For the ordinary user from the technical field, the learning effort (for a customized product) is too high. Here, fast planning and a quick rendition of the data for a presentation are all that is asked for. The extra effort for creating a game-like environment will, in most cases, neither be financially rewarded, nor will it be met with understanding, in the technical environment. If ever a client should come up with such a demand, it is worthwhile to look in the yellow pages, where it should not be too difficult to find somebody specializing in this field. But, as mentioned before, if you like experimenting in the field of visualization, everything you learn will contribute to the quality and acceptance of all projects at hand.

---

[3] www.anark.com

[4] www.quest3d.com

[5] www.virtools.com

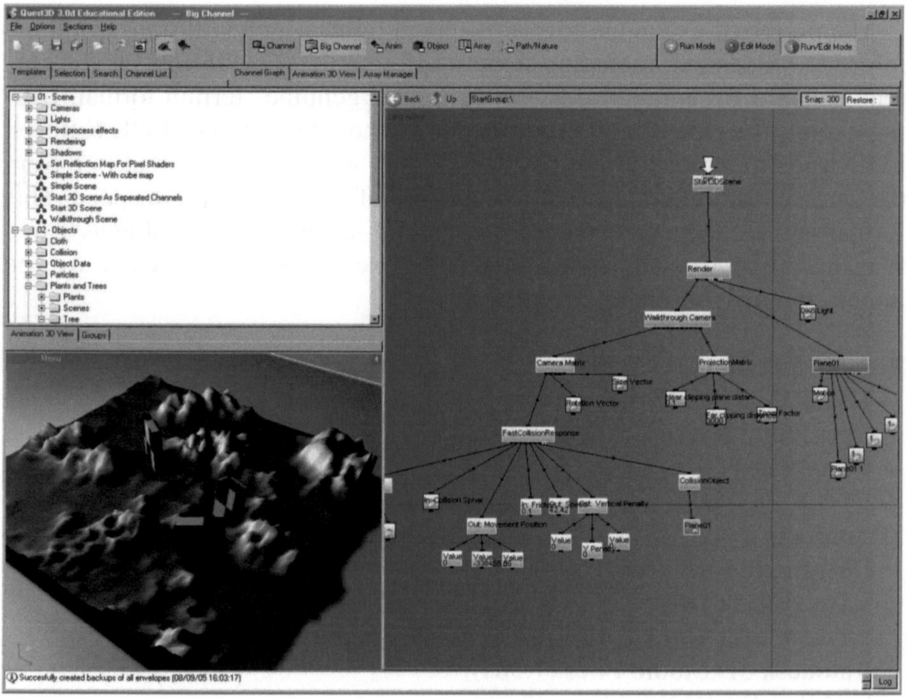


**Fig. 255.** Screenshot of the Quest3D environment

## Terrain Affairs

Let's get back to the visualization of pure terrain data. Only few tools are able to handle large amounts of geometrical and texture data. The effort involved in getting familiar with these tools is usually great.

In the foreground of the interactive visualization is the requirement to represent original data without optimization and with as simple a navigation as possible. On the commercial market TerrainView[6] is one of the few tools specialized in terrain visualization (planning and GIS data). This program is an editor and viewer which is very easy to use and which has a few extras. It will even allow the visualization of the whole of Switzerland in a 2,0 meter grid web streamed, thanks to sophisticated algorithms.

[6] www.viewtec.ch

The preferred exchange format is OpenFlight, which will transfer all the information needed, including texture information. OpenFlight[7] is the current standard for simulation systems; it is a "genuine" terrain format. During export, the OpenFlight program allows the integration of LOD settings. LOD leads to a reduction of the data density with increasing distance from the camera, for the graphic edition, and therefore an increase in speed. This format is included as a standard in most 3D formats, or available as a free Plug-In. Apart from this, there are more exchange formats which can be imported into TerrainView:

**Terrain Formats (Geo Data)**
TerraPage Files (*.txp)
OpenFlight Files (*.flt))
VTree Files (*.vt)

**3D Objects**
OpenFlight Files (*.flt))
VTree Files (*.vt)
VTree Files (*.vtc)
Autodesk 3D Studio Files (*.3ds)
VRML 1.0/VRML97 Files (*.wrl)
Wavefront Files (*.obj)
LightWave Object Files (*.lwo)
Design Workshop Files (*.dw)
Geo Files (*.geo)
OSG Files (*.osg).

Basically, there are two types of 3 dimensional data:
Terrain – pure terrain data, normally available with geo-reference
Objects – any 3D object, which can be imported into the scene.

There can be any number of 3D objects in a scene, but there is usually only one terrain. The assignment of the materials is done automatically, and the model is available for a walkthrough, a flythrough, or any kind of viewing. Whenever desired, any 3D objects can be added to the scene, blended in or out, or be manipulated separately. The handling is self-explanatory and similar to a flight simulator. In short, it is easy and totally uncomplicated.

---

[7] OpenFlight is an open coded data format. If you want to protect your geo data, there is, for TerrainView, the IVC converter for coding and optimising the data.

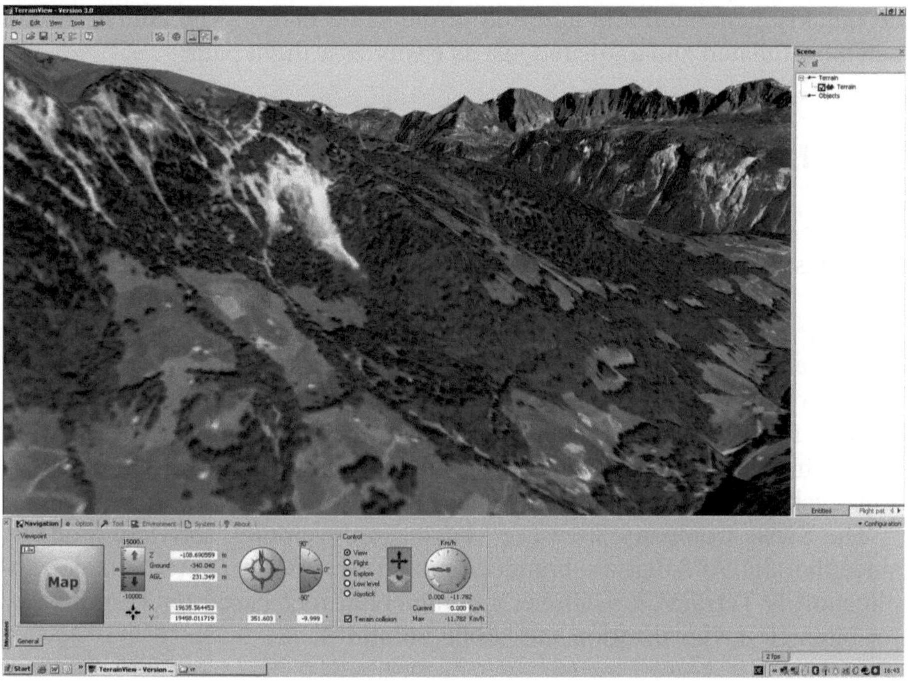


**Fig. 256.** Screenshot of the user interface

All the necessary navigation components like compass, height control, and speed indicator can be found in the lower third of the monitor. The respective height of the camera can be set parametrically or with the help of a scale (e.g. the speed in km/h at which the camera is moving across the terrain). Thus the values are precisely specified, and not given in system units. This is interesting, particularly when working with "real data", because it allows the accurate assessment of the duration of flights or drives.

Another useful option is the fact that a 2D graphic can be loaded as an overview. This can be set up or geo-referenced via the coordinates for the 3D scene, and will interact with the represented 3D scenario. This means that the selection of a spot on the 2D overview will automatically call up the corresponding 3D view.

There are a few gimmicks and some useful tools hidden in the modules of the program. Among others, in the area "tools", there is the option of defining points of interest (POI). These are corresponding to separate camera positions, and can be defined, selected and activated as link targets.

The particular condition of the weather, sunny and clear, for instance, or cloudy, is included as an environment parameter. It goes without saying that the shape and density of the clouds as well as background pictures can be quickly generated and adjusted in the same dialog. The course of day can be simulated and animated just as easily as a flight path.

## Virtual Globe

Another very interesting method to view 3 dimensional data has come up in the last few years. It is called Virtual Globe.

ESRI[8] launched ArcGlobe in mid-2003 and Google[9] its Google Earth in 2005. Leica Geosystems[10] from Hexagon put ERDAS Imagine and Virtual Explorer software on the market. Intergraph[11] has a product called Image Viewer for MicroStation. Autodesk[12] has acquired sophisticated software for high quality 3D modeling, visualizations and animations, such as Autodesk Civil3D, 3ds Max or Maya.

The Swiss Company ViewTec[13] has been developing software products since the late 90´s that are branded under TerrainView$^{TM}$. Their major application is TerrainView-Globe, a commercial tool based on TerrainView for interactive visualization and Globe Solutions.

Upstream core technologies required for the realization of applications with Virtual Globes are

Remote Sensing and imagery acquisition (satellite, aerial, terrestrial)
Image processing, ortho rectification and projection
Geographical and other spatial information processing
Global positioning systems (satellites, receivers, emitters)
3D Modeling Tools to create additional information and include designs and constructions

[8] http://www.esri.com
[9] http://www.google.com
[10] http://www.leica-geosystems.com/
[11] http://www.intergraph.com/
[12] http://autodesk.com
[13] http://www.viewtec.ch

A ***Virtual Globe*** can be described as a system representing geo information and content with regards to one reference (e.g. WGS 84[14]), including imagery, digital maps, elevation and 3D models, vector data, weather data, and data from Real Time GPS or surveillance systems (satellites, planes, helicopters, UAV[15]s, cameras).

Virtual Globe applications provide functionality for accurate representation of the earth up to a very high level of detail, for 3D visualization and navigation.

Virtual Globes can represent and combine geo referenced data from different sources and systems:

***Remote Sensing*** is in the broadest sense the measurement or acquisition of information of an object or phenomenon by a recording device that is not in physical or intimate contact with the object. In practice, remote sensing is the utilization at a distance (as from aircraft, spacecraft, satellite, or ship) of any device for gathering information about the environment.

A ***geographic information system*** (GIS) is a system for creating, storing, retrieving, analyzing, displaying and sharing geographically referenced spatial data and associated attributes.

***Cartographic modeling*** refers to a process where several thematic layers of the same area are produced, processed, and analyzed. Dana Tomlin (1990) used raster layers but the overlay method can be used more generally. Operations on map layers can be combined into algorithms, and eventually into simulation or optimization models.

Digital cartography can be considered a (semi) automated process of producing maps from a GIS system. The representation of GIS output through a digital map is considered equivalent to the stage of visualization.

The main factors in the creation of an interactive Virtual Globe solution can be described as follows:

**3D Globe Geo Content** – a Virtual Globe solution has to be able to represent the most important types of geo information such as elevation information, imagery, vector delineations and surfaces and annotations.

**3D Models** - additional 3D models are important in order to create a convincing Virtual Globe solution (entire cities and textured buildings, landmarks, plants and production facilities, power lines, road, railway and pipeline networks, street culture, vegetation, dynamic 3D objects such as planes, cars, people).

---

14 http://www.wgs84.com/

15 UAV - Unmanned Aerial Vehicle

**Navigational Information** – for navigation, Points of Interest, flight paths and so called Hot Spots are important.

**Navigation** – the navigation has to be simple, easy and instinctive. If it is a Flythrough or a Walkthrough or just a "Find Target", the user has to be able to get familiar with the system quickly. Using the mouse or a joystick to navigate in an intuitive way is one of the most effective ways of getting the solution accepted.

**Web Based Application** - Internet revolutionized the way we communicate with each other and share knowledge and information. Therefore the hosting and streaming of 3D Geo Data, Terrain Models and additional information inside a web-based environment is a basic requirement.

**Webcams, Intelligence and Remote Control** - spatial representation of moving or stable objects in 3D on the ground or in the air through the help of intelligence and reconnaissance technology (satellite, UAV, Webcam data streams, GPS) is a very important necessity. While the Network Centric Intelligence Technology has its origins in defense and domestic security, civil applications can include endangered species, wild animals surveillance, or live event broadcasting.

**Presentation and communication** - presentation and communication of complex data for the public is an important detail. Any type of Globe-related enterprise or project to plan and implement (e.g. engineering, urban land use planning, architecture, network infrastructure, navigation, security, defense, United Nations missions) needs adequate presentation and communication platforms to various audiences.

**Large Screen Presentation** - in professional use, larger screen dimensions are needed for command and control (launch of the space shuttle, air control, and more) as well as in the fields of virtual reality, flight simulation and entertainment, where multi-channel and multi-projection screens allow immersive visualizations and experiences with virtual 3D landscapes.

The most well known applications are
Viewtec TerrainView-Globe[16]
Google Earth[17]
Microsoft MSN Virtual Earth[18]
NASA and World Wind[19]
WW2D[20]

---

[16] http://www.viewtec.ch
[17] http://earth.google.com
[18] http://local.live.com
[19] http://worldwind.arc.nasa.gov/
[20] http://ww2d.csoft.net/

## Summary

Interaction is the "cherry on top" of 3D visualization. However, it is a complex subject, which, in most cases, will entail considerable effort and additional cost.

There are various types of interaction. Presented here are the most current procedures, which are well accepted in the environment of the PC, and can be pursued in the course of regular project work.

There are some general requirements and techniques for generating interactive presentations. Among these are basic requirements for the particular geometry of terrain models, standards for editing materials and textures, and of course, some preconditions for navigation with data to be viewed interactively.

The problems resulting from too many modifiers as well as the problem of texture baking were highlighted.

One has to distinguish between a very limited interaction, like for instance the use of a panorama taken over from a 3D application, and the complex rendition of geometrical data in Real Time applications.

The most popular methods free of charge are:

QuickTime VR for panorama representations
VRML for the rendition of geometrical data

Of the commercial products, programs like Anark, Quest3D, and Virtools were touched upon but not discussed in much detail.

The program TerrainView is a very easy to use tool for interactive viewing of terrain data.

Planning, simulation and monitoring in a computer generated realistic 3D world will increase the demand for Virtual Globe applications.

Initiatives like Internet2, with a high-performance backbone network that operates at up to 100 Gbits per second[21], will dramatically increase usage and exchange of massive amount of data for 3D visualization.

Collaborative virtual environments (CVE), the underlying concept used in Virtual Globes will connect people and projects across the whole world.

[21] UCAID http://www.internet2.edu

# Practical Examples

*We conclude this book by taking a look at two practical projects with strong purchase for visualization.*

## Workflow of a digital landscape design

In the context of a research project, the HSR University for Technology in Rapperswil examined the question, how "Earth Grading by real time GPS" as part of an entirely digital work routine to tape a planned golf course could work.

### Reason and Planning

The new 9-hole golf course in Bad Ragaz, which was planned by the well-known golf course architect Peter Harradine, is embedded into the delightful landscape of the Swiss Rhine Valley. By sensitive handling of the area which includes a large water surface, the project fits well into the environment. For the first time, a digital work routine was demonstratively run through from beginning to end in a HSR research project for gardening and landscape gardening in Switzerland. Using GPS, a measurement office collected the data for the golf course.

The office of Peter Harradine designed a digital working plan (e.g. contour maps and the planting plan, among others) using 2D CAD. During the design phase, further terrain and vegetation data, which turned out to be very important, were surveyed by the planners directly using the GS20.

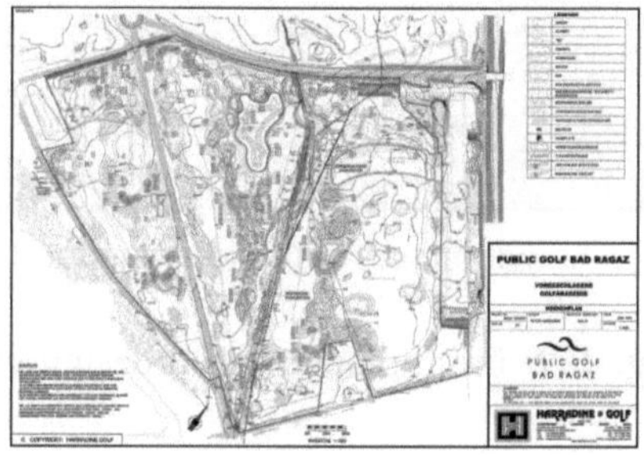


**Fig. 257.** Contour Map

Leica Geosystems developed this GPS especially for those who have no experience in surveying. With the measurement data and the contour map a 3D terrain model was generated. The terrain model was integrated into the DEM 25 basic model with ortho photos.

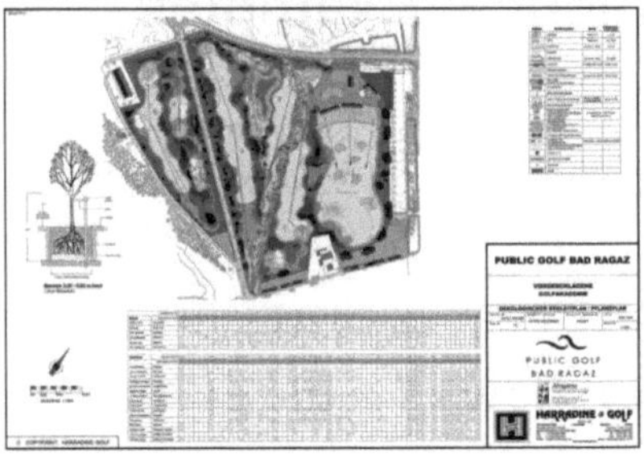

**Fig. 258.** Planting plan

## DTM and Visualization

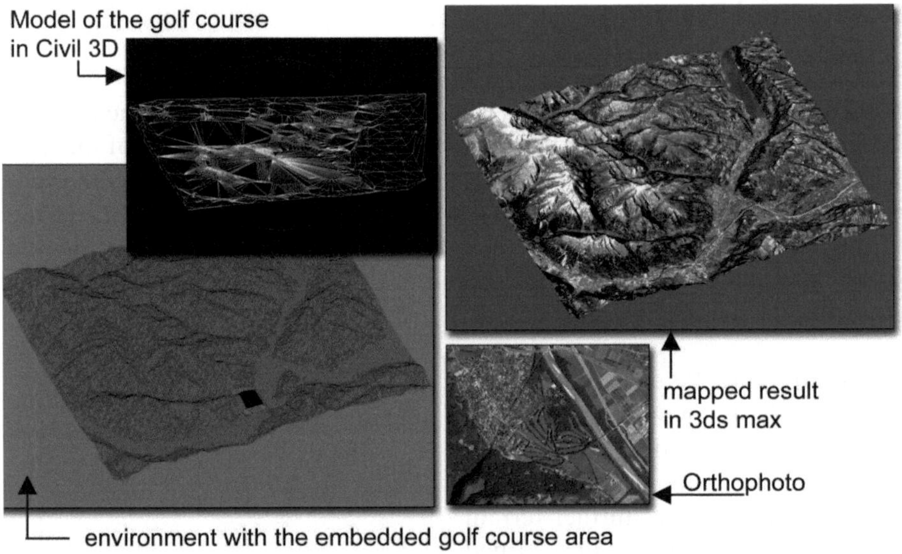


**Fig. 259.** From planning and design of the 3D model to the integration into the environment

By optimizing the interface between the programs 3ds max and TerrainView, the data for an interactive walkthrough of the planned situation was available.

## Feasibility

After the landscape architects had made modifications based on the 3D real time walkthrough, the landscape modeling took place using a 3D GPS bulldozer system.

## Construction

**Fig. 260.** Using the Leica GPS machine automation system and thereby directly transferring the DTM data into practice.

The geodetic points were directly transferred from Autodesk Civil 3D to the Leica GPS machine automation system.

The painstaking work of pegging out the terrain was avoided.

In the Fall of 2005, a civil engineering company modeled the public golf course using a 3D GPS bulldozer. Usually, normal shovel excavators are used in garden and landscape architecture.

In the meantime, Leica Geosystems has developed a new version of the 3D GPS bulldozer system. With this, it is now possible to realize even complicated terrain models, such as those created in garden and landscape architecture, from start to finish without having to peg out the terrain, while achieving a high level of accuracy.

## Software Used

Planning and DTM - AutoDesk Civil 3D 2005
GISDataPro – Leica Geosystems
GIS Data preparation and conversion - ArcGIS
3D visualization and animation - AutoDesk Media 3ds max
3D Interaction – TerrainView

## Conclusion

The technical possibilities are vast. The connection between the collection of data, planning and execution is excellent. The visualization of data is a decisive factor in the feasibility of a planned measure. In the project at hand, being able to take a look at the data in the form of a DTM helped immensely to make the whole project run smoothly.

It is possible to say that GPS automatic machines can be used outside of their classical fields of application, such as mining and civil engineering. The appropriate technologies and state of the art are in existence and are already supported by the machines which are used on construction sites today.

However, this is only under condition that the planners give their plans to the building contractors in the form of 3 dimensional data sets, based on the elevation model created by the surveyor of the landscape in question!

## Design for the Federal Horticultural Show

For the planning of the horticultural show in Munich in 2005, Rainer Schmidt's company placed an emphasis on the visualization of the plans in order to communicate the landscape architectural elements clearly and efficiently ahead of the show.

An excerpt from the main idea:
*The existing landscape park of the BUGA (BundesGartenschau), with its generous paths, lawns and groves, forms an important point of attraction and is a tranquil contrast to the neighboring exhibition spaces.*
*The diverse microstructures of park life become a central part of the horticultural show in a subtle way: photographs taken of life in the park are exhibited in the so-called "houses of perspective", thereby giving the visitor a new perspective.*
*The temporary horticultural competitive exhibitions will take place in clearly structured areas of the future building. These BUGA exhibition spaces, as well as several gardening themes within the permanent exhibit of the landscape park follow basic biological forms (cell garden, leaf garden and garden of powers). By means of reduction, the temporary exhibition spaces become expressive and memorable landscapes, in which a change in one's point of view or perspective can be experienced. The linear terrace forms an important connecting line in this endeavor; it is the joint between constructed reality and fictional imagination.*

## Planning

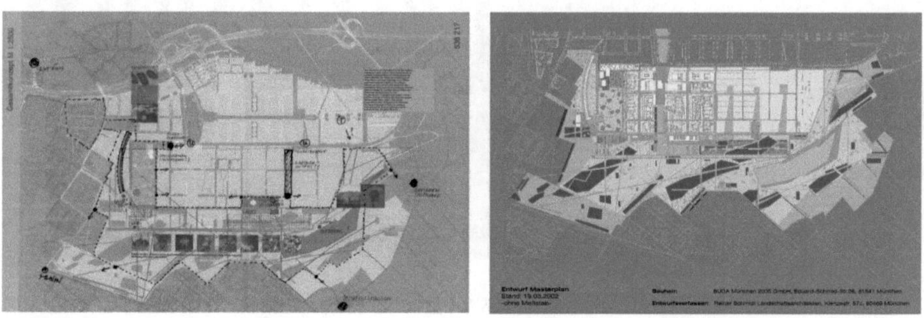

**Fig. 261.** The picture on the left shows the integration of the landscape park and the surrounding area into the concept of the horticultural show. A change in perspective is the theme which is followed through in all aspects of the Bundesgartenschau. The picture on the right shows Riem as a calm contrast to the neighboring exhibition spaces.

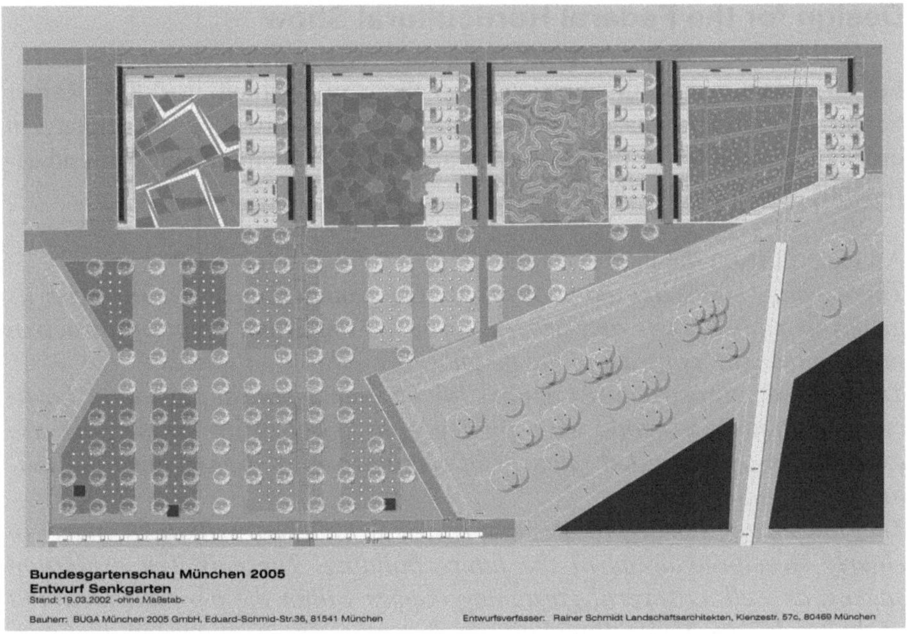


**Fig. 262.** Layout of the gardens

## Examples of Visualizations

**Fig. 263.** 3D visualization of the gardens to the power of $10^{-4}$, Epidermis of the underside of the leaf of the marsh marigold

**Fig. 264.** 3D visualization of the gardens from a bird's eye perspective

All images courtesy of Rainer Schmidt Landscape Architects.


| | |
|---|---|
| Name of the project: | Bundesgartenschau 2005 München |
| Contact person: | Prof. Rainer Schmidt |
| Telephone number: | 0049 - 89 – 20 25 35 –0 |
| Email: | mail@schmidt-landschaftsarchitekten.de |
| Homepage: | www.schmidt-landschaftsarchitekten.de |
| Address: | Klenzestrasse 57c, D-80469 München |


## Software Used

Planning and concept: VectorWorks 11.0, AutoCad 2000
Image editing: Photoshop 7.0
3D visualization: 3ds max

# Glossary

*A list of the most important terms used in 3D visualization*

## Terms and Definitions

### 123

| | |
|---|---|
| **2.5 D** | 3D surfaces, where every X, Y point has only one Z value (e.g. terrain models without overhangs or caves). |
| **3DS-Format** | File format of 3D Studio Release 4 |

### A

| | |
|---|---|
| **A/D- Converter (DAC)** | Converts digital input signals into analog output signals, i.e. visual data in the display memory of the graphics card is converted to video signals, so that it can be shown on the monitor. |
| **Additive colors** | When a white surface is illuminated by several light sources in different colors, the result is an additive color. The (additive) mixing of two complementary colors adds up to the color white. The additive mixing of colors is the basis of computer graphics. |
| **ADI** | Abbreviation of Autodesk Device Interface, an interface to the products of Autodesk |
| **Aliasing** | Staircase effect |
| **Alpha Blending** | In addition to the values for the colors red, green, and blue, a transparency value is assigned to every pixel. This method allows the representation of materials in various degrees of transparency, e.g. normal glass, milky glass, fog, smoke. |
| **Alpha Channel** | The color rendering of pixels is made up of the colors red, green, and blue. By adding another byte, the transparency of the pixel can be defined - this additional byte is called the "Alpha Channel". In an 8bit Alpha Channel, 256 shades of transparency can be represented. In image editing, this channel is often used for saving masks. |
| **Ambient light** | The ambient light defines the basic brightness and the basic coloring of your scene. |

| | |
|---|---|
| **Analog** | Continuously varying electronic signal for generating data. Opposite: digital. |
| **ANIMATICS** | Animated scripts, normally generated by integrating separately drawn up scribbles in the form of a QuickTime or an AVI movie. The finished video sequences are called ANIMATICS. |
| **ANSI** | American National Standards Institute |
| **Anti-Aliasing** | This is the interpolation of the colors of neighboring pixels, to prevent "pixel visibility" in a picture. Anti-Aliasing is normally used to avoid the "staircase effect" in diagonal edges and lines. |
| **Area Light** | In the field of studio photography, this method is often used for simulating natural light. Area lights generate very smooth transitions between light and dark |
| **ASCII Arc Info Grid Format** | A spatial data model defined by a raster of even sized pixels (gray scale picture). An attribute value, e.g. for defining the elevation, is assigned to every pixel. For editing, the extension Spatial Analyst or 3D Analyst, is needed. |
| **ASCII** | American Standard Code for Information Interchange. This is a simple code for the digital saving of alphanumerical data, which can be read by nearly every computer system; it is made up of a maximum of 256 signs. |
| **Aspect Ratio** | The Aspect Ratio indicates the proportions of a still frame or a movie frame as the ratio of width to height. As a rule, the aspect ratio is either given as the quotient of width and height (e.g. 4:3), or as the ratio number on the basis of 1 (e.g. 1,333). |
| **ATKIS** | Acronym for "Amtliches topographisch-kartographisches Informationssystem", the German official topographical-cartographical information system. |
| **Atmospheric effects** | Mist and fog in a nature scene, the light veil of haziness in the distance, cloud formations and smoke, are all examples of atmospheric effects. |
| **Attribute** | Information linked to objects in a GIS or CAD system describing geometrical or subject characteristics (e.g. area, volume). |
| **AVI-Files** | Short for Audio Video Interlaced, the video format of Microsoft |

# B

| | |
|---|---|
| **Batch** | In a batch job, certain programs or commands are executed automatically by the computer, without further interference |

by the user.

**Bezier Curve** Bezier curves will provide the model with softer forms than straight-lined polygons. The course of the Bezier curves is interpolated by selecting surface points in regular intervals. The degree of curvature is defined by tangents positioned along the curve.

**Beziér-Spline, B-Spline** B-Splines are an extension of Bezier curves. The term spline comes from the flexible spline devices used by ship-builders and draftsmen to draw smooth shapes.

**Billboard** Square transparent plane containing a bitmap for the representation of a 3D object. In landscape visualization, a frequently used method for representing the vegetation.

**Bitmap** Digital raster picture

**Bitmap - Image Format** Saving of a graphic representation by dividing up the graphic into even and regular picture elements. Some various possibilities for use the word "Bitmap" are:

Bitmap - general term for pixel graphics

Bitmap - *.bmp – picture format of Microsoft (see bmp)

Bitmap - value for color depth (2 colors: black and white)

**Bits per pixel** Number of bits representing the color information of a pixel.

8 Bit corresponds to 256 colors

16 Bit will give you around 65.000 colors (High Color)

with 24 Bit, 16,7 million colors (True Color) can be represented.

32 Bit per pixel will allow the representation of 16,7 million colors plus an 8 Bit Alpha Channel for transparency information (see Alpha Channel).

**Blinn Shading** A special method for shading based on Phong shading, the significant difference being that highlights on glossy surfaces will have a more rounded shape.

**Blur** The effect in a picture or in a movie, that moving objects will appear blurry. This effect can either be directly assigned to specific objects, as object characteristics, or it can be assigned as soft focus to the entire scene, in the rendering dialog.

**BMP files** Bitmap. Windows file format for pixel graphics.

**Boolean Modeling** With this method, by using the logic operators AND, OR, NOT, objects can be added or subtracted, and their intersection can be determined.

**Bump Mapping** To provide a texture with a near-reality structure, it has to be superimposed by a bump map which will then transfer the elevation data to the texture. A Bump Map is a gray scale picture where different values for brightness are used to indicate different topographical levels of height. The darker the values of the Bump Map, the more depth will be added to the texture.

# C

**CAD** Computer Aided Design

**CAM** Computer Aided Manufacturing

**CAVE** Computer Augmented Visualization Environment

**CGA** This is an abbreviation for Color Graphics Adaptor by IBM, one of the first standards for color graphics. It can either represent 320x200 pixels with four colors, or 640x200 pixels with two colors.

**Chrominance** Chrominance is part of a video signal linked to the color value and containing information on color shade and saturation. This color component basically increases the brightness and luminance of a color picture

**Cinepak** Cinepak is used for compression of 24 Bit videos for CDs. It is available on Windows as well as on Macintosh computers. Best results are obtained when the Cinepak-Codec is applied to the pure original data which have not yet undergone a very lossy compression. Cinepak is a very asymmetrical codec, i.e. the decompression by Cinepak is much faster than the compression.

**Clipping** All currently invisible areas of a 3D picture (depending on the perspective to be calculated) are left aside and ignored during later picture editing.

**Codec** The encoder/decoder is a piece of hardware for the conversion of analog and digital audio and video signals. The term is also used for hardware or software that can compress and decompress audio or video data (compression/ decompression), or for the combination of encoder/decoder and compression/decompression.

**Color Depth** Also called pixel depth. Number of bits per pixel. A system using 8 bit per pixel, can represent 256 colors. A system using 16 bit per pixel can represent 65.536 colors. A system using 24 bit per pixel can represent more than 16.7 million colors. 24 bit colors are often called Real Color Representation, because the human eye can distinguish between only around 6 million different color shades, i.e. a lower number than is available in a 24 bit color system. 24 bit means 8 bit for every RGB. In a pixel depth of 32 bit, the additional 8 bit are used for the Alpha Channel.

**Constant Shading** Method for constant shading. Every surface of an object is calculated and represented as being even. This method is very similar to Flat Shading, but includes a number of added highlights.

**Control Points** Control vertex points for editing Splines or NURBS, see CV.

**Coordinates** A system of coordinates is used for spatial orientation. The precise position of a point within a 3 dimensional space is determined by the values of X, Y, and Z.

**CPU** Abbreviation for Central Processing Unit, the main processing chip of the computer, e.g. Pentium chip.

**CV** Abbreviation for Control Vertex

# D

**DDS Format** The Data Dictionary System allows the exchange of parameter definitions and binary data between software programs and the user

**Delaunay Triangulation** A method of connecting arbitrary sets of points together in a network of triangles which meet certain mathematical criteria (specifically, the circle described by the three points in any triangle contains no other point in the set), used in creating a TIN

**Delta Image** This is a picture containing only data which has changed since the last picture. Delta images are a very efficient tool for the compression of visual data.

**DEM** Digital Elevation Model (*.DEM – NASA-Format)

**DEM Base Model** The Swiss base model includes digitalized contours, as well as the main alpine tectonic faults as polylines, and an irregularly distributed number of points. It is available at swisstopo (Bundesamt für Landestopografie), for Switzerland.

**DEM Matrix Model** The DEM Matrix Model was interpolated from the DEM Base Model. It has a standardized grid and a regular distribution of points.

**Depth Cueing** Depth cueing plays an important role in the realistic representation of 3D models: objects in the distance will appear blurrier and darker than objects nearby. This effect is achieved by fading with black pixels. In other words, depth cueing is a kind of black haze.

**Diffuse Color** This is the color of a directly illuminated surface. When asked about the color of an object, it is the diffuse color that is normally given.

**Digital** (1) Method for the representation of sound or other waves as a succession of binary signals. (2) Method of radio setting where the desired frequency is set on digital. (3) Numerical representation of information data. Opposite: analog.

**Digitize** Translation of an analog signal in digital data, e.g. scanning of an image.

**Digitizer** Translation of an analog signal into digital data, e.g. when scanning a picture.

**Distant Light**

**Dithering** Method for representing pictures with originally high color depth, in good quality but with less depth of color , and thus requiring a smaller file size. The picture is rasterized and the color values are interpolated.

**DOM** "Digitales Oberflächenmodell" - digital surface model, representing the surface of the earth, including vegetation and buildings.

**Double Buffering** Also called page flipping. While one picture is shown on the monitor, the computation of the next picture is in progress. It is saved in a special memory and is shown only after the computation has been completed. This way, a visible construction, line by line, is avoided, which reduces the flickering of the picture in animations, games and video replay.

**DPI** Dots per Inch. Measure for the resolution precision of a digital representation.

**DTM** Digital Terrain Model (without buildings or vegetation)

**Dummy** A dummy is an object used as an aid in animation. A dummy is not rendered during the computation of a picture. Dummies are a popular device for the animation of separate limbs and for links between objects.

**DXF POLYMESH, DXF POLYFACE** These are special polylines. DXF POLYMESH can be compared to a wireframe model, whereas DXF POLYFACE defines a surface. Because in POLYFACE the indices of all corner points are included, the datasize is about 2.2 times larger than that of POLYMESH.

**DXF Files** Data Exchange Format, corresponding to the AutoCAD original format DWG as ASCII set of data. The DXF format is NOT a standard and changes with every new release of AutoCAD.

# E

**ECD** Short for Enhanced Color Display, by IBM, for a resolution of 640 x 350.

**Environment** The environment of 3D models. There is a choice of real environments (universe, sky, clouds), or surreal environments (anything the imagination allows).

**Environment Map** In this map, the calculation of the radiation of the reflected environment is simulated, by using a bitmap as reflection. In this way, an existing picture can be used for reflection as well as for refraction.

**EPS-Files** Encapsulated Postscript. EPS is an extended version of Postscript. Apart from the option of editing vector and pixel information, it also includes the option of using release paths.

**Extrusion** With this method, depth of space is added to a 2 dimensional model, along one of the spatial axes.

# F

**Field of View (FOV)** The focal length determines the field of view, i.e. those areas of a 3D scene which are to be rendered

**Fields (upper, lower)** The picture on the television screen actually consists of two pictures (fields), alternating 50 times per second. This means that instead of a linear sequence of 25 single frames, 50 half frames are chasing across the TV screen. During rendition, a line jumping procedure is used, in which a "normal" full picture is divided up into specific lines, i.e. into even and uneven lines. The uneven lines are reserved for the first half picture, the even lines make up the second half picture (field, semi picture). This method was originally developed to reduce the bandwidth for the transmission of television signals.

**Filter** The special effects in a video clip or picture can be modified by a filter. Filters are also used for correcting color contrast, brightness, or balance.

**Flare** Simulation of a light refraction, generated by a bright ray of light hitting the lens of a camera.

**Flat Shading** "Flat" shading method. All surfaces of an object are represented by using only one color , i.e. only one color value per surface. The objects computed by flat shading have a sharp edged appearance.

| | |
|---|---|
| **Focal Length** | The focal length is the main information of the camera lens. As a rule, objectives cover a specific, fixed focal distance: 28 mm, 50 mm, 85 mm.<br>There are, however, zoom objectives covering a specific range of focal distances: 20 – 28 mm, 28 – 85 mm, 70 – 210 mm. |
| **Fog** | A fading effect, depending on the distance of the object to the viewer. |
| **Forward moving kinematics (FK)** | The hierarchical linking of an object on a higher level to an object on a lower level (downward kinematics). |
| **FOV, Field of View** | The area which you can see. |
| **FPS** | Frames per second. Measuring unit of picture ratio in videos and animations. A picture ratio of about 20 frames per second will result in a continuous sequence of pictures. Television is broadcasted in 25 pictures per second. |
| **Fractal** | Fractal geometry is based on the principle of "self-similarity". Fractal objects are made up of elements showing the same structure as the original elements on the higher level of hierarchy. Benoit Mandelbrot is a mathematician who has done intensive studies in this field. |
| **Frame** | Single video picture |

# G

| | |
|---|---|
| **G-Buffer** | In video post production, picture layer actions use G-Buffer masks, instead of RGB and Alpha masks. These are based on graphic puffer channels. |
| **Geo Referencing** | Assignment of a coordinate system to objects. |
| **Geo Tif** | Pixel-based data format containing geo-referenced information. |
| **Ghosting** | Formerly, animations were drawn on a transparent celluloid foil, through which one could see the previous hand drawn frames. |
| **GIF Files** | Graphics Interchange Format. GIF is a LZW compressed format, which was developed to reduce the file sizes and transmission times via telephone as much as possible. |
| **GIS** | Geographic Information System. A system for the retrieval, administration, analysis and representation of huge amounts of spatial data including their thematic attributes. |
| **Glow** | Illumination effect generated during rendering. |

**Golden Section** The “golden ratio” is the division of a specified length in such a way that the ratio of the entire distance to the larger part is equal to the ratio of the larger part to the smaller part. According to this theorem, the division of a length or a surface in a ratio of 3:5 will appear harmonious to the viewer.

**Gouraud Shading** Optimized type of Flat Shading, in which the edges are interpolated with intermediate color values, resulting in a picture that is less irritating to the viewer.

**GPS** Global Positioning System

**Gray Scale** A gray scale picture is made up only of shades of gray. There are normally 254 different shades of gray, plus black and white: thus there are altogether 256 shades.

**Grid format DTM** In digital terrain models, the grid format is the composition on the basis of an even matrix used for the definition of the terrain model in rows and columns.

**Grid** A mesh forming the basic structure of a raster representation.

# H

**HLS** Hue, Lightness, Saturation. Color system for the definition of colors according to depth of color , brightness, and saturation.

**HSDS** Hierarchical SubDivision Surfaces is the subsequent subdivision of surfaces, with the goal of obtaining a higher resolution for an improved quality representation.

# I

**IGES** Initial Graphics Exchange Specification. ANSI standard for the definition of a neutral format for the data exchange between different CAD (Computer-Aided Design), CAM (Computer-Aided Manufacturing), and computer visualization systems.

**IK, Inverse Kinematics** In contrast to the real-life course of the movement of, for instance, the human arm, where the chain of movement (kinematics) when lifting the arm, will start at the shoulder, and then move down to the upper arm, then to the lower arm, and finally to the hand, 3D models are easier to control from the end of the chain of movement. This reverse movement control is called Inverse Kinematics.

**Indexed Color** Indexed color pictures include a table of colors in their data. This table lists all colors that can occur in the picture. For an indexed 16 bit color picture the table contains 16 listed colors (4 Bit), for an indexed 256 color picture there are 256 colors (8 Bit). Further colors can be simulated, like in the gray scales of a purely black and white representation, by positioning pixels of different colors close to each other. The eye will then see colors that are not actually present in the color table. You can change pictures into indexed color pictures, if you want to be able to load them in some programs like Windows Paintbrush, or to show them on a monitor which can only represent 256 or 16 colors.

**INTEL Indeo Video** For the compression of 24 bit video for CDs. Like the Cinepak Codec, the INTEL Indeo Video achieves higher compression rates, a better picture quality, as well as a replay speed than the Microsoft Video 1 Codec, and it is available for Windows as well as for Macintosh computers.

**Interactive** Special modus of the operating system where data input and program control by the user can be done during program execution

**Interface** An connection between two or more components of a system.

**Interlaced Representation** The screen is divided into lines. In the interlaced method, first all even, then all uneven lines are built up for the picture on the monitor. This method allows a higher graphics resolution, but the monitor will flicker more than a non-interlaced monitor, where the entire monitor surface with all its lines is actualized every time.

**INTERLIS** INTERLIS is a descriptive and transfer mechanism for geo data. With this universal language, the experts can model their data precisely in order to provide software applications and interface services. The basic idea behind INTERLIS is that a digital exchange of structured information is only possible when the institutions participating in the exchange have a precise and identical idea about the characteristics of data to be exchanged. Further information under www.interlis.ch.

## J

**JPEG, JPG Files** Joint Photographic Experts Group. JPEG is the current format for the representation of photographs and other semi/halftone /continuous tone pictures in HTML files in the World Wide Web and other online services.

# K

**Keyframe** A keyframe is a basic picture used for comparison with other frames in order to detect differences. Keyframes are used for defining various animation sequences.

**Keyframe Animation** This term originates from the world of animated cartoons drawn by hand, and was used for the keyframes of the movie which were drawn by the main designer. The frames in between the keyframes were done by numerous "drawing servants".

# L

**Landsat Mosaic** A Landsat Mosaic is a satellite picture of a large area, e.g. Switzerland, in a resolution of 25 m, composed without visible transitions, by a mosaic of several geocoded and radiometrically fitted scenes. The satellite pictures were taken by the remote sensing satellite Landsat 5, from a height of 705 km. The picture comprises only the spectral channels 3, 2, and 1 and is available in the resolutions of 25m and 100m, in TIFF format.

**Landscape Models** Landscape models represent the objects of a landscape in a flexible vector format. They are made up of subject layers (e.g. a traffic system). Every layer includes geo referenced objects as point, line, or surfaces. Attributes and relationships are assigned to every object (topology).

**LandXML** This format represents the topology of a TIN in a list of knots and elements. Land XML is an OpenSource format supported by many GIS products and software producers.

**Lathe, Rotation Object** 2-dimensional vector graphics are changed into a 3-dimensional form, by turning them around an axis. The semi-sectional cut of any object is an ideal basis for a Lathe model.

**Lens Effects** Lens effects are illumination effects as perceived by the human eye. Just imagine you were looking directly into the sun (never ever do this without a filter, please). You would see many rays and a diffuse illumination around the sun, a strongly accentuated aura. In computer graphics, lens effects are usually generated via filters during post production.

**Light Decrease, Attenuation** With increasing distance to its origin, the intensity of light decreases. Objects near the source of light will appear

| | |
|---|---|
| | lighter than objects further away from it. |
| **Lights** | In 3D programs there are usually several types of light or illumination sources. |
| **LOD** | Level Of Detail – Real-time objects have to be represented in different levels of detail, normally with less detail in the distance and with more detail in the foreground. |
| **L-System** | Descriptive system for the simulation of the development of graphic structures, used mainly for the generation of pictures showing plants. |
| **Luminance** | Part of a video signal determining the degree of brightness - generally the scale of black and white underlying a color picture. |

## M

| | |
|---|---|
| **Mapping** | Mapping is the current term used for assigning a texture,(e.g.) a material, to a 3D object. |
| **Material** | The term material describes the sum of all surface characteristics of an object. |
| **Matte Object** | By using matte objects, invisibility is assigned to specific objects. They are able to cast shadows, but you cannot see them. |
| **Meshes** | Description of objects, usually by a polygonal mesh. |
| **Metadata** | Data about data, like source, date, precision, and further attributes. |
| **Metal Shading** | Metal shading is used when dealing with strongly reflecting surfaces like metal or glass. |
| **Microsoft Video 1** | Compression of analog video - a lossy spatial compression supporting color depths of 8 to 16 bit. |
| **MIP Maps** | MIP maps are a collection of optimized bitmaps (4x4, 16x16, 256x256 pixels provided in addition to the main texture. MIP maps are used especially in real time applications, to increase flight speed. The abbreviation is taken from the Latin “multum in parvo”, meaning “a lot in a small space”. |
| **Morphing** | Special effect in which one form is slowly transformed into another form. |
| **MOV files** | Movie, the Apple data format for audio and video |
| **MPEG4** | MPEG4 is an internationally standardized method for the memory-saving recording of moving pictures including |

| | |
|---|---|
| | multi-channel sound. This format is supported by the "Motion Joint Picture Expert Group" (MJPEG). An important characteristic of all JPEG formats so far is the upward compatibility, i.e. updated encoders/decoders (Codecs) continue to accept older formats of the same edition. |

# N

| | |
|---|---|
| **Normal** | A surface normal, or just normal to a flat surface is a three-dimensional vector which is perpendicular to that surface. A texture is normally mapped on the side containing the surface normal. |
| **NTSC** | National Television Standards Committee, the video standard used in North America, large parts of Central and South America, and in Japan. |
| **NURBS** | Non-Uniform-Rational-B-Spline. NURBS are precisely defined mathematical functions, the precision of which does not depend on the detailing of elements, unlike polygonal modeling. |
| **NURMS** | Non-Uniform-Rational-Mesh-Smooth, a special Mesh-Smooth-Modifier integrated in Max. |

# O

| | |
|---|---|
| **OBJ-Files** | Alias Wavefront data format |
| **Omni-Light** | Like a light bulb, an omni light emits its light evenly in all directions. It is not possible to specify the focus of the rays of this light type. |
| **Opacity** | Light impenetrability. Low values correspond to a high transparency, high values to a low transparency. |
| **OpenFlight** | Data format of MultiGen Paradigm, which has become standard in the field of simulation of terrain data. |
| **OpenGL** | 3D software interface (3D API) for Windows NT and Windows 95, licensed by Microsoft and based on Iris GL from Silicon Graphics. |
| **Ortho photo** | An ortho photo is an aerial or satellite picture which has been corrected by geometrical transformation, and corresponds to an orthogonal projection of a terrain onto a cartographic surface. |
| **Ortho Rectification** | Method of fitting a photograph to a costant horizontal scale. |

**Overshoot** Option for the illumination of the entire scene, independent of the actual cone of light. However, shadows will only be cast in the area of the cone of light.

# P

**PAL** "Phase Alternate Line". PAL is the television standard of most European countries.

**Particle System** Snow, rain, dust etc. can be simulated by particle systems.

**Patch** Patch objects are suitable for generating slightly curved surfaces.

**PDF-Files** The Portable Document Format (PDF), like HTML, is a platform-independent file format for the administration of text, vector and picture data.

**Perspective** A view based on the way the human eye sees. Objects in the distance are shown to be smaller, which gives the impression of spatial depth.

**Phong Shading** In Phong shading, the edges and surfaces are smoothed. Highlights of evenly glossy surfaces are realistically rendered.

**Photometry** Photometry is the simulation of the distribution of light in a defined environment, based on physical factors.

**Pivot Point** A pivot represents the local centre and the local coordinate system of an object.

**Pixel Shading (Dither)** Representation of a color by mixing closely related colors.

**Pixel** Abbreviation of "picture cell", the smallest unit represented on the screen, also called pel, "picture element". Pixels can be compared to the dots that make up photographic representations in newspapers.

**POI, Point of Interest** Main subject of the representation. The entire scene revolves around the POI.

**Polygon** A surface consisting of any number of straight lines. The smallest polygon is a triangle.

**POV, Point of View** Position of the camera. Point from which the scene is viewed.

**Primitive** Objects consisting of simple geometric forms, like spheres, cubes and pyramids. Since these forms can be easily described in mathematical terms, they save on computing time and memory space. In modeling, one should always fall back on these basic objects, as long as this remains realistic.

**Procedural Map** In contrast to the fixed matrix of a bitmap, a procedure map is generated with the help of mathematical algorithms. A great variety of forms can be generated by using procedure maps

**Projection** Mathematical formula for the conversion of points of a sphere (e.g. planet earth) onto a plane (e.g. for a plan).

**PSD File** Photoshop file format

**PS File** Postscript. Actually a controllable printers' language.

# Q

**QuickTime** QuickTime Movie. Apple file format for audio and video

# R

**Radiosity** In radiosity, light is considered to be energy, which allows the physically nearly correct computation of the diffuse distribution of light in a defined space.

**Raytracing** In Raytracing, a "virtual" projection surface is set up between the eye of the viewer, i.e. your camera, and the scene to be viewed. This projection surface corresponds to the desired resolution of the resulting planned picture, with respect to length and width.

**Real time** In real time 3D visualization, the tedious computation of the pictures and scenes becomes redundant, because one can make use of highly specialized hardware (high-tech graphic cards) which is nowadays included in most standard PC systems. These graphic cards include many algorithms in the hardware which would normally have to be dealt with by the software and the CPU of the computer. With the help of these graphic cards, computations in real time can be done, i.e. with more than 25 frames per second.

**Refraction, Index of R.** The degree of refraction which occurs when light encounters a more or less transparent surface. A virtual sphere of glass will look very realistic when the refraction index of the "glass texture" is set equal to the typical refraction factor of real glass.

**Refresh Rate** Number of pictures represented per time unit. Software videos have a fixed refresh rate. During replay the actually shown rate of pictures may differ by a large degree from that existing on the tape (see also fps).

**Render** This is the computation procedure which is necessary to transform a 3D model or a 3D scene into a 2D representation. This procedure can be done by various computational methods, each requiring different calculation efforts and resulting in a different quality of the finished result.

**Resolution** Number of horizontal and vertical pixels on the screen. The higher the resolution, the higher the precision of the picture.

**RGB 8 Colors** The RGB 8 color data type is a 3 bit type, in which every pixel can take on one of eight colors. The RGB 8 color pictures are automatically changed to indexed 16-color pictures; the eight original colors are retained while space is provided for eight additional colors. However, it is not possible to change another data type into the RGB 8 color type.

**RGB Color space** By additive mixing of the colors red, green and blue, a picture with an infinite number of colors can be represented on the monitor. That is why the editing of visual data is done with the data from the RGB file. The three color vectors are forming a color environment, where the value for the color black is positioned/set in the origin, and the value for the color white is found at the opposite end.

**RGB Format** In most cases, a TIF file is too large to be used in a real time environment. The picture needs to be optimized for quick loading by a graphics card; a suitable format for this is .rgb. RGB files can be written from Photoshop, when a Plug-In has been installed. The Plug-In for Photoshop can be found under: http://www.telegraphics.com.au

**RGB True Color (True Color)** RGB is short for Red Green Blue. In this data type, the colors are composed by mixing a specific percentage of each of these three basic colors. The percentage of each of the three colors can vary in 256 grades. By mixing these color grades, 16.7 million possible color combinations can be obtained (3 times 8 bit = 24 bit, 2 to the power of 24 = 16.7 million). The human eye is not capable of distinguishing between such an immense number of color grades. Hence the term True Color = representation in real-life colors.

**RLA Files** RLA is a widely used SGI format. The RLA format supports 16 Bit RGB files by a single Alpha Channel. RLA is an excellent format for further editing of 3D visualizations, because it is possible to save depth information in this format.

**Rotoscopy** A rotoscopy is the method of importing video frames as a background for suitable objects.

**RPF Files** RLA files are replaced by RPF files as the preferred format for the rendering of animations which require further editing or additional work on specific effects.

# S

**Saturation** Saturation defines the depth of a color . A color with a high degree of saturation is very intense, a color with a low degree of saturation will look faded.

**Scanline Renderer** A Scanline Renderer is used for calculating the brightness of every single pixel of a surface. This ensures a realistic transition from bright to dark, as well as the positioning of highlights and textures.

**Scene** The sum of all elements of a 3D composition (models, lights, textures, etc.).

**Self Illumination** By self illumination you obtain the effect of an illuminated surface which does not cast a shadow. This illusion is achieved by replacing the shadows on the surface with diffuse colors.

**Shading (Flat, constant, Phong, Blinn)** Shading, or rendering, allows the definition of the colors on a warped surface, for giving the object a natural look. In order to achieve this, the surfaces are divided up into small triangles.

**Shadow Color** Generally, the color of the shadow will correspond to the complementary color of the main light source.

**Shadow Map** The bitmap generated by the renderer during the first rendering of a scene is called the shadow map. In a Scanline Renderer, the generation of the shadows is done by so-called shadow maps. The precision of the shadow is defined by the size of the shadow map. The higher the value, the more precise is the calculation of the shadow.

**Shape file** A vector data format for saving the position and other geographical attributes.

**Skinning** With this technique you put a skin around the cross beams of a model.

**SMPTE** SMPTE (Society of Motion Picture and Television Engineers) is a time setting used in most professsional animation productions. The SMPTE format indicates minutes, seconds, and frames from left to right, separated by a colon, for example 01:14:12 (1 minute, 14 seconds, and 12 frames).

**Spline** A curve which is defined by control points outside the curve. The method is similar to the Bezier curves, but is defined by a different mathematical algorithm. The term spline was originally used in ship building, where metal tapes around the body of the vessel were bent into the required shape by weights attached to specific points.

**Spotlight** Spotlights are basically point lights, the difference being that the light distribution is limited to the area within a defined cone, a 360° spotlight being a point light. In most 3D programs this angle can be set on a continuous scale.

**SPOT Mosaic** Spot Mosaic is the new 5 m satellite mosaic picture of Switzerland composed by several geo-coded and radiometrically adjusted scenes. The satellite pictures were taken by the remote sensing satellite Spot 5, from a height of 822 kilometers. The true color picture with a resolution of 5 m was composed by integrating two mosaic pictures taken simultaneously at two different resolutions, and will be available in TIFF format. http://www.npoc.ch

**Subtractive Color** This is the color which results when one or more colors of the incoming light are absorbed. When all the colors of the spectre are absorbed, we perceive the color black.

**Sunlight** In sunlight, the rays of light reach the scene in a parallel fashion. Depending on the time of day and the constellation of the weather, the angle of the incoming light, as well as the brightness and the coloring (from a high percentage of white at midday to increasing shades of red in the evening) will vary. This, in turn, has an influence on the intensity, the direction and the coloring of the shadows of the illuminated object

# T

**Texture Mapping** The representation of a bitmap on an object, taking into account the adjustment of perspective (e.g. with respect to the pattern of the wallpaper, the wood grain on furniture).

**Textures** By mapping a 3D model by a texture with an organic-looking surface, the 3D model will get an appearance closer to reality. A texture can be a bitmap picture, or a procedurally generated map (see procedure map). The appearance and the illumination of the texture can be adjusted by parameters like refraction or transparency.

**TGA Files** Truevision format. TGA was developed for use with systems working with Truevision video cards.

**TIF(F) Files** Tagged-Image File Format. TIF is a flexible Bitmap format, supported by nearly every software for painting, picture editing, and page layout. TIF pictures can be produced on nearly all desktop scanners.

**Tiles** The patterns of floor tiles or wallpaper are typical examples for the use of tiles. A small segment, i.e. a tile, is used and repeated n times, depending on the pattern. In 3D visualization usually a specific number of tiles in U and V direction of the map will be given. U and V are the local axes of the respective map.

**TIN** Triangulated Irregular Network. In each case the nearest neighboring points are combined into irregular triangles; the surfaces obtained in this way will form the terrain model.

**Topologic Data Structure** The method for saving graphic data in such a way that the topological relations between the different objects can be calculated.

**Topology** The science of the position and arrangement of geometrical bodies within a specified space.

**Transformation Matrix** Linear algebra is the language of 3D graphics. All transformations within a scene, e.g. scaling, rotation, or positioning of objects, are defined by 4 x 4 transformation matrices.

**Transparency** Transparency is the characteristic which allows the light to penetrate a material, in contrast to opacity, which prevents the light from penetrating.

**True Color Representation** Simultaneous representation of 16.7 million colors (24 or 32 bits per pixel). The color information saved in the display memory is directly transferred to the D/A changer, without having to proceed through a translation table. Therefore the color data for each pixel has to be saved separately. True Color representation is based on the fact that the human eye cannot distinguish between more than 16.7 million colors.

## U

**UVW Coordinates** Mapping coordinates are called UV or UVW coordinates. These letters refer to the spatial coordinates of an object, in contrast to the XYZ coordinates used for describing the entire scene.

## V

**Vector Graphic** Saving of graphic data based on the coordinates of single points, or specific lengths of geometric curves.

**Vertex** Point

**VGA** Short for Video Graphics Adaptor by IBM, with a standard resolution of 640 x 480 pixel and 16 colors.

**Virtual Reality** VR is a term used for describing interaction in 3D worlds. The most advanced example for this are CAVE technologies.

**Volume Model** The digital definition of a geometrical object including its 3 dimensional characteristics.

**VRAM** Short for Video Random Access Memory; memory chip for fast graphic charts.

**VRML** Virtual Reality Markup Language,

## W

**Wireframe Model** The skeleton structure of a 3D model, consisting either of polygons, or Bezier curves, or even NURBS (see below).

**Wireframe** Wireframe view of a 3D model. The skeletal body of the model without textures. There are wireframe views with hidden lines, and meshwire views including all lines (transparent model).

**World Coordinate System** The world is the universal coordinate system for all objects in a scene.

## X

**XML** The Extensible Markup Language is a standard for the generation of machine readable as well as human readable documents, structured like a tree.

**Xref** External reference. An external file referred to.

## Y

**YUV Color space** Visual data of single pictures are composed by one part brightness and two parts color . The color values are obtained by differentiation with the value for brightness. This method was originally used in color television technology

## Z

**Z Buffer** Information about the 3D depth (position in the 3rd dimension) of every pixel. Z Buffering is a method for removing hidden surfaces.

# Figures and Tables

## Index of Figures

## Index of Tables

# Literature

**Bishop/Lange** (2005): Visualization in Landscape and Environmental Planning. Taylor & Francis, London.
**Buhmann/Paar/Bishop/Lange** (2005): Trends in Real-Time Landscape Visualization and Participation. Proceedings at Anhalt University of Applied Sciences, Wichmann, Heidelberg.
**Buhmann/von Haaren/Miller** (2004): Trends in Online Landscape Architecture. Proceedings at Anhalt University of Applied Sciences, Wichmann, Heidelberg.
**Buhmann/Ervin** (2003): Trends in Landscape Modeling. Proceedings at Anhalt University of Applied Sciences, Wichmann, Heidelberg.
**Buhmann/Nothelf/Pietsch** (2002): Trends in GIS and Virtualization in Environmental Planning and Design. Proceedings at Anhalt University of Applied Sciences, Wichmann, Heidelberg.
**Coors/Zipf** (2005): 3D-GEOINFORMATIONSSYSTEME, Wichmann Verlag, Heidelberg.
**Deussen, Oliver** (2003): Computergenerierte Pflanzen. Springer-Verlag, Heidelberg.
**Draper, Pete** (2004): Deconstructing the Elements with 3ds max 6, ELESVIER Oxford.
**Ervin/Hasbrouck** (2001): Landscape Modeling. Digital Techniques for Landscape Visualization, McGRAW-HILL, New York.
**Fleming, Bill** (1999): Advanced 3D Photorealism Techniques. Wiley Computer Publishing.
**Grant, C.** (2006), Library Systems in the Age of the Web, 27p. http://www.nelinet.net/edserv/conf/cataloging/google/grant.ppt#1
**Gugerli, David, Hsg**. (1999): Vermessene Landschaften - Kulturgeschichte und technische Praxis im 19. und 20. Jahrhundert - Interferenzen I - Chronos Verlag Zürich
**Hehl-Lange, Sigrid** (2001): GIS-gestützte Habitatmodellierung und 3D-Visualisierung räumlich-funktionaler Beziehungen in der Landschaft - ORL-Bericht 108/2001, ORL ETHZ Zürich.
**Hochstöger Franz**, (1989): Ein Beitrag zur Anwendung und Visualisierung Digitaler Geländemodelle. Dissertation Technische Universität Wien.
**Imhof, Eduard** (1982): Cartographic Relief Presentation. De Gruyter, Berlin

**Lange, Eckart** (1998):Realität und computergestützte visuelle Simulation. Eine empirische Untersuchung über den Realitätsgrad virtueller Landschaften am Beispiel des Talraums Brunnen / Schwyz. Dissertation ETH Zürich.

**Mach, Rüdiger** (2000): 3D Visualisierung. Galileo Press, Bonn.

**Mach, Rüdiger** (2003): 3DS Max 5. Galileo Press, Bonn.

**Maguire/Goodchild/Rhind** (1991): Geographical Information Systems (Vol.1 Principles / Vol.2 Applications), Longman Scientific & Technical, Harlow.

**Miller, C.L. & Laflamme, R.A.** (1958): The Digital Terrain Model - Theory & Application in: Photogrammetric Engineering, Vol. XXIV, No. 3, June 1958. The American Society of Photogrammetry.

**Muhar, Andreas** (1992): EDV-Anwendungen in Landschaftsplanung und Freiraumgestaltung. Verlag Eugen Ulmer, Stuttgart.

**Petschek, Peter** (2005): Projektbericht KTI Forschungsprojekt gps rt 3d p – gps und echtzeitbasierte 3D Planung. HSR Hochschule für Technik Rapperswil, Abteilung Landschaftsarchitektur, Rapperswil.

**Petschek, Peter** (2003): Projektbericht KTI Forschungsprojekt Planung des öffentlichen Raumes - der Einsatz von neuen Medien und 3D-Visualisierungen am Beispiel des Entwicklungsgebietes Zürich-Leutschenbach. HSR Hochschule für Technik Rapperswil, Abteilung Landschaftsarchitektur, Rapperswil.

**Sheppard, Stephen** (1989): Visual Simulation. A User's Guide for Architects, Engineers, and Planners, Van Nostrand Reinhold, New York.

**Westort, Caroline** (2001): Digital Earth Moving. First International Symposium, DEM 2001, Manno, Switzerland, September 2001, Proceedings. Springer, Heidelberg.

## Used Software

| | |
|---|---|
| Operating System: | Microsoft: Windows XP Pro<br>www.microsoft.de |
| Word Processing: | Microsoft: Word 2000/XP<br>www.microsoft.de |
| Picture Editing: | Adobe: Photoshop 7.01<br>www.adobe.de |
| Vector Graphic: | Corel: CorelDraw 12<br>www.corel.de |
| Landscape Modeling: | AutoDesk: Civil 3D<br>www.autodesk.de |

| | |
|---|---|
| GIS-Applications: | ESRI: ArcGIS 9.1<br>www.esri.com |
| 3D-Visualization: | AutoDesk Media: 3ds max 7.5<br>www.autodesk.de<br>Itoo Software: Forestpack<br>www.itoosoft.com<br>Digital Elements: Worldbuilder 4<br>www.digital-element.com<br>Planetside: Terragen<br>www.planetside.co.uk<br>Eon Software: Vue 5 Infinite<br>http://www.e-onsoftware.com |
| Plants | Digital Elements: Verdant<br>www.digital-element.com |
| Interactive Applications | Anark Corporation: Anark Studio 3.0<br>www.anark.com<br>Act3D: Quest 3D 3.0<br>www.quest3d.com<br>Viewtec: TerrainView 3.0<br>www.viewtec.ch |

# Index

MIX
Papier aus verantwortungsvollen Quellen
Paper from responsible sources
FSC® C105338

If you have any concerns about our products,
you can contact us on
**ProductSafety@springernature.com**

In case Publisher is established outside the EU,
the EU authorized representative is:
**Springer Nature Customer Service Center GmbH**
**Europaplatz 3, 69115 Heidelberg, Germany**

Printed by Libri Plureos GmbH
in Hamburg, Germany

you have any concerns about our products
you can contact us on
ProductSafety@springernature.com

Publisher is established in the EU
the EU authorised representative is
Springer Nature Customer Service Center GmbH
Europaplatz 3, 69115 Heidelberg, Germany

Printed by [illegible] GmbH
in Hamburg, Germany

MIX
Papier aus verantwortungsvollen Quellen
Paper from responsible sources
FSC® C105338

If you have any concerns about our products,
you can contact us on
**ProductSafety@springernature.com**

In case Publisher is established outside the EU,
the EU authorized representative is:
**Springer Nature Customer Service Center GmbH**
**Europaplatz 3, 69115 Heidelberg, Germany**

Printed by Libri Plureos GmbH
in Hamburg, Germany